Aprovecho Institute

Charcoal: Small Scale Production and Use

A Publication of
Deutsches Zentrum für Entwicklungstechnologien – GATE
in: Deutsche Gesellschaft für Technische Zusammenarbeit (GTZ) GmbH

Friedr. Vieweg & Sohn   Braunschweig/Wiesbaden

APROVECHO is a small non-profit association of people from several countries, based in Oregon, USA. They offer ordinary people help in handling the inevitable changes that come with dwindling resources. Locally, they teach and practice techniques for simple living: housing, cooking, heating, small-scale food production. In developing countries, their role is that of facilitator. They help people to create and to adopt technologies that use their own skills and resources.

Written by Michael Boutette and G. Edward Karch, illustrated by Michael Boutette

CIP-Kurztitelaufnahme der Deutschen Bibliothek

Boutette, Michael:
Charcoal: small scale production and use :
a publ. of Dt. Zentrum für Entwicklungstechno-
logien – GATE in: Dt. Ges. für Techn. Zusammen-
arbeit (GTZ) GmbH / [written by Michael Boutette
and G. Edward Karch]. Aprovecho Inst. [Ill. by
Michael Boutette]. – Braunschweig ; Wiesbaden :
Vieweg, 1984.
  ISBN 3-528-02009-1
NE: Karch, G. Edward:; HST

Published by Friedr. Vieweg & Sohn Verlagsgesellschaft mbH, Braunschweig
Printed in the Federal Republic of Germany by Lengericher Handelsdruckerei, Lengerich

ISBN 3-528-02009-1

Contents

Preface

Charcoal has been used as a cooking fuel for thousands of years. Its advantages such as smokelessness and light weight have long been recognized. As awareness of fuelwood shortages increases, people are taking a new look at charcoal.

This booklet presents an introduction into various aspects of smallscale charcoal production, such as economics of charcoal production and use, charcoal making devices, kiln operation and improvement of stove efficiency. It will help the extension worker to get involved in charcoal production, but it does not pretend to give all information available on charcoal. No book can replace practical experience, and the best method for the development worker to get experience is to work with the indigenous smallscale charcoal makers.

Construction of small kilns is described in more detail in the GATE-brochure "Construction of small kiln systems".

We would like to thank those people who helped in the creation of this book: Elisabeth Gern and Elaine Hogg for editing and consultation, Anita Ochoa for unending patience, Laquetta Karch for final edit and Aprovecho for consistent support. Their help and encouragement made this book possible.

Thank You.

Michael Boutette & Ed Karch

1. Economics and Efficiency of Charcoal

Charcoal production is a process in which heat is applied to organic matter inside a relatively air-tight container commonly referred to as a kiln. The charcoal resulting from this process has twice the potential energy per unit weight as air-dried wood. However, 70 to 80 percent of the energy in the wood was used to produce the charcoal. How can a fuel which requires that much energy to produce fit into a plan aimed at energy conservation and saving natural resources? To begin, let's take a look at the fuelwood situation as it presently stands.

Fuelwood Crisis

Most of the people of the world cook and heat with fuelwood or charcoal. As the world's population has increased, so has the demand for fuelwood. With the increasing need for land to produce food and provide living space there has been a decrease in the area of land left for growing trees. Unregulated cutting of trees for fuel and for charcoal production has worsened the problem in many areas. Cyclic weather patterns, soil loss and degradation and concentration of population in urban areas where charcoal is most commonly used have further contributed to the problem of supplying fuel.

Past Problems with Charcoal

In the past, charcoal production has commonly been based on low yielding, unregulated production practices which often result in rapid forest destruction. This problem is magnified when you consider that charcoal is mainly used in urban areas. Increased urbanization in developing countries can cause increased charcoal use and accelerate deforestation.

In an attempt to solve the deforestation problem, several countries have made efforts which range from laws abolishing charcoal production entirely, to various regulatory

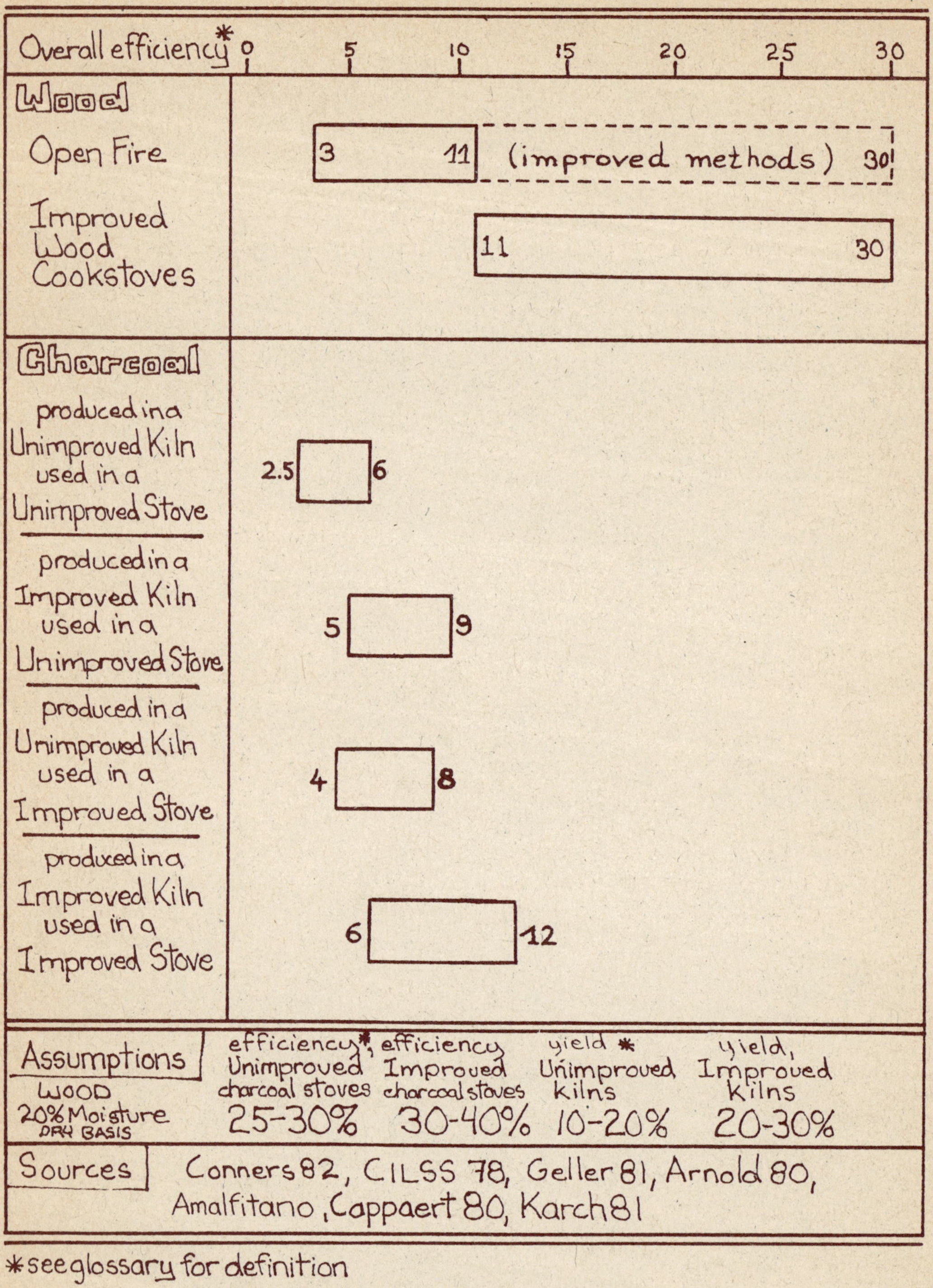

Overall efficiency *

| | 0 | 5 | 10 | 15 | 20 | 25 | 30 |

Wood

Open Fire: 3 — 11 (improved methods) 30

Improved Wood Cookstoves: 11 — 30

Charcoal

produced in a Unimproved Kiln used in a Unimproved Stove: 2.5 — 6

produced in a Improved Kiln used in a Unimproved Stove: 5 — 9

produced in a Unimproved Kiln used in a Improved Stove: 4 — 8

produced in a Improved Kiln used in a Improved Stove: 6 — 12

Assumptions	efficiency * Unimproved charcoal stoves	efficiency Improved charcoal stoves	yield * Unimproved kilns	yield, Improved kilns
WOOD 20% Moisture DRY BASIS	25-30%	30-40%	10-20%	20-30%

Sources	Conners 82, CILSS 78, Geller 81, Arnold 80, Amalfitano, Cappaert 80, Karch 81

* see glossary for definition

Fig. 1-1: Comparing the Overall Efficiences of Wood and Charcoal

practices attempting to establish some form of control over charcoal making. Examples of these controls include requiring a permit for cutting wood, producing, transporting or selling charcoal, and price controls. The purpose of most of these restrictions is to inhibit charcoal use and extend fuelwood resources. However, such rules and regulations are difficult to enforce as long as there is a demand for charcoal.

Advantages of Charcoal

If charcoal presents such problems, why is it still in use? In reality, charcoal is rarely used in rural areas. It is in towns and cities that the advantages of charcoal over wood are the most obvious:

1. needs less storage space;
2. can be used in small, efficient, inexpensive stoves;
3. doesn't deteriorate in storage:
4. doesn't have to be cut up;
5. contains more energy for the same weight;
6. is cheaper to transport;
7. burns cleanly, resulting in little or no smoke in the kitchen as well as less air pollution in the town.

A New Perspective on the Efficiency of Charcoal

The advantages of charcoal as an urban cooking fuel will result in an increasing demand in many regions of developing countries, and a substitution of charcoal by kerosene and gas is unrealistic in respect to the low income situation of the population. So every effort has to be made to improve the efficiency of charcoal production and use.

We will see that this can result in economic advantages of charcoal, compared with fuelwood.

If various sources of energy are to be compared for their efficiency, the entire process of production, transportation, and use must be considered and not just the production process. Let's take a closer look at the relationship between charcoal and wood using a graph derived from Table A in Appendix A (Fig. 1-1).

We can see from this chart that there is virtually no difference in overall efficiency when comparing open fires and improved charcoal technology. As improvements are made on charcoal kilns and stoves, charcoal appears a more and more attractive option. It must be seen, however, that improved charcoal production and use cannot be as effective in saving energy as improved wood stoves.

Economic Advantages of Charcoal

As you will see in the following tables, it is in transportation where charcoal has the most marked advantage over fuelwood, due to charcoal's higher energy content per unit weight. A truck loaded with charcoal can carry twice the energy of one loaded with wood (Fig. 1-2).

The tables in this chapter apply only to the conditions stated in each table. To obtain

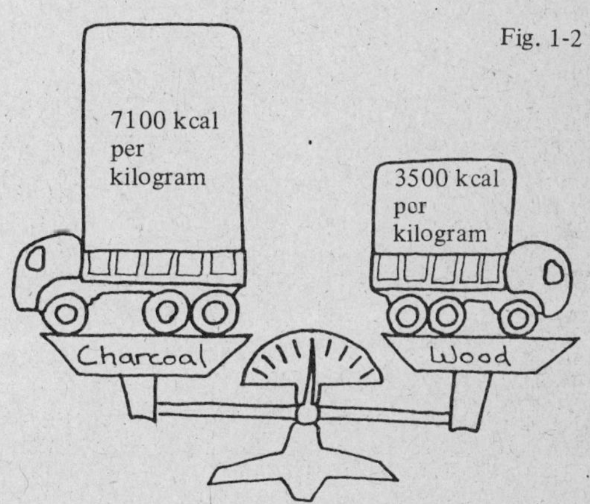

Fig. 1-2

figures for your own area the following information is needed:

1. cost of wood to the consumer;
2. cost of charcoal to the consumer;
3. transportation costs per tonne by whatever means are normally used in your area;
4. production costs of wood and charcoal — these can be obtained by subtracting transportation costs and any other intermediate costs from the cost of the fuel to the consumer;
5. cost of any other common fuels;
6. the efficiencies of stoves used in your area. To calculate this figure you will need the VITA manual, *Stove Testing Procedures* (see bibliography).

The chart in Fig. 1-3 shows that, for the conditions stated, using wood is more expensive than using charcoal if the haul is over 95 km. Clearly, for longer distances it is more cost effective to make the wood into charcoal and then haul it.

The major factor affecting the hauling of wood is the moisture content. A higher moisture content in the wood not only reduces the amount of energy available for use, since energy must be used to evaporate the water, it also increases the amount of weight hauled per volume.

A Comparison of Charcoal, Wood and Other Fuels

Let us now expand our view and compare the cost of charcoal and fuelwood to that of other fuels, using an example from a West African country in 1980.

Table 1 shows kerosene to be the best buy, and this is probably the most common fuel

Fig. 1-3: Economics of Transport, Wood vs. Charcoal (derived from Table B, Appendix A)

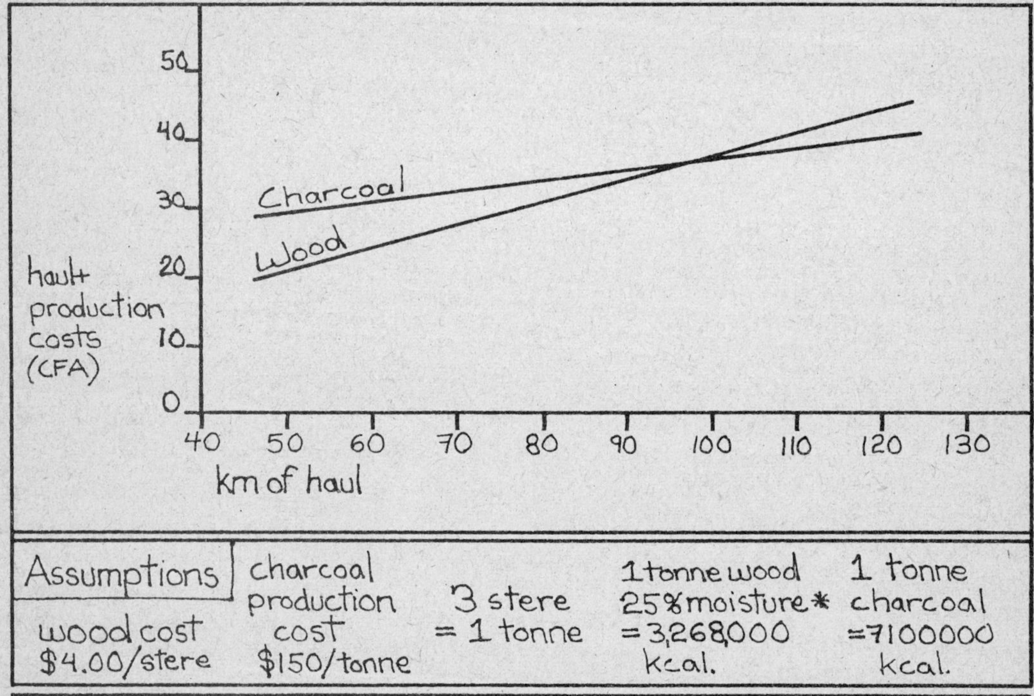

8

in the cities of this particular country. This, however, does not represent a free market situation as kerosene is subsidized by the government.

In Table 2 we can see that wood or charcoal would be the best buy. Gas is very seldom available in rural areas and was included only for comparison. Charcoal is also seldom

Table 1: Comparative Fuel Costs for the Consumer: Urban, Traditional Stoves

fuel	unit	cost (CFA franc)	energy (1000 kcal units)	cost per 1000 kcal	% stove efficiency	cost at the pot per 1000 kcal
kerosene	liter	60	8.32/liter	7.21	40	18.03←
gas	15 kg tank	4500	11.5/kg	26.09	70	37.27
charcoal	35 kg sack	1500	7.1/kg	6.04	25	24.14
wood	350 kg stere	2000	1200/stere	1.67	7	23.85

Table 2: Comparative Fuel Costs for the Consumer: Rural, Traditional Stoves

fuel	unit	cost (CFA franc)	energy (1000 kcal units)	cost per 1000 kcal	% stove efficiency	cost at the pot per 1000 kcal
kerosene	liter	60	8.32/liter	7.21	40	18.03
gas	15 kg tank	5500	11.5/kg	31.88	70	45.55
charcoal	35 kg sack	750	7.1/kg	3.02	25	12.07←
wood	350 kg stere	750	1200/stere	.63	7	9.00←

Table 3: Comparative Fuel Costs for the Consumer: Urban, Improved Stove Efficiencies

fuel	unit	cost (CFA franc)	energy (1000 kcal units)	cost per 1000 kcal	% stove efficiency	cost at the pot per 1000 kcal
kerosene	liter	60	8.32/liter	7.21	40	18.03
gas	15 kg tank	4500	11.5/kg	26.09	70	37.27
charcoal	35 kg sack	1500	7.1/kg	6.04	40←	15.10←
wood	350 kg stere	2000	1200/stere	1.67	30←	5.50←

available in a rural village and seldom used, even by charcoal makers themselves. Except in a wood shortage area or for a special use such as blacksmithing, special food preparation, or clothes ironing, charcoal's rural role is as an income generator.

In Table 3 we can see the effects of improving stove efficiency. Here wood becomes by far the best buy, being 9.60 CFA/1000 kcal cheaper than charcoal. This is a rather large difference to pay, but the advantages noted earlier may be worth it.

In the rural improved stove table (Table 4) the cost spread between wood and charcoal is not as wide but the cost is lower overall for fuelwood. Here the convenience of charcoal is probably within economic reach except that cash surplus is usually non-existent in villages, and the cost for wood is usually measured in the time spent to gather it. By comparison, charcoal is a cash commodity and, as pointed out above, is very seldom used in rural areas. From this, and the advantages noted earlier, we can see that charcoal is more appropriate for the urban user.

Tables 5 through 8 show how improving wood and charcoal production practices and stove performance affects costs to the urban consumer.

Table 4: Comparative Fuel Costs for the Consumer: Rural, Improved Stove Efficiencies

fuel	unit	cost (CFA francs)	energy (1000 kcal units)	cost per 1000 kcal	% stove efficiency	cost at the pot per 1000 kcal
kerosene	liter	60	8.32/liter	7.21	40	18.03
gas	15 kg tank	5500	11.5/kg	31.88	70	45.55
charcoal	35 kg sack	750	7.1/kg	3.02	40	7.55
wood	350 kg stere	750	1200/stere	.63	30	2.10

Table 5: Comparative Fuel Costs for the Consumer: Urban, Charcoal Production Improvements, No Change in Stove Efficiencies

fuel	unit	cost (CFA francs)	energy (1000 kcal units)	cost per 1000 kcal	% stove efficiency	cost at the pot per 1000 kcal
kerosene	liter	60	8.32/liter	7.21	40	18.03
gas	15 kg tank	4500	11.5/kg	26.09	70	37.27
charcoal	35 kg sack	850	7.1/kg	3.42	25	13.68
wood	350 kg stere	2000	1200/stere	1.67	7	23.86

Table 6: Comparative Fuel Costs for the Consumer: Urban, Charcoal and Wood Production Improvements

fuel	unit	cost (CFA franc)	energy (1000 Kcal units)	cost per 1000 kcal	% stove efficiency	cost at the pot per 1000 Kcal
kerosene	liter	60	8.32/liter	7.21	40	18.03
gas	15 kg tank	4500	11.5/kg	26.09	70	37.27
charcoal	35kg sack	850←	7.1/kg	3.42	25	13.68←
wood	350kg stere	1200←	1200/stere	1.00	7	14.29←

Table 7: Comparative Fuel Costs for the Consumer: Urban, Improved Stoves and Charcoal Production

fuel	unit	cost (CFA franc)	energy (1000 Kcal units)	cost per 1000 kcal	% stove efficiency	cost at the pot per 1000 Kcal
kerosene	liter	60	8.32/liter	7.21	40	18.03
gas	15 kg tank	4500	11.5/kg	26.09	70	37.27
charcoal	35kg sack	850←	7.1/kg	3.42	40←	8.55←
wood	350kg stere	2000	1200/stere	1.67	30←	5.57←

Table 8: Comparative Fuel Costs for the Consumer: Urban, Stove, Charcoal and Wood Production Improvements

fuel	unit	cost (CFA franc)	energy (1000 Kcal units)	cost per 1000 kcal	% stove efficiency	cost at the pot per 1000 Kcal
kerosene	liter	60	8.32/liter	7.21	40	18.03
gas	15 kg tank	4500	11.5/kg	26.09	70	37.27
charcoal	35kg sack	850←	7.1/kg	3.42	40←	8.55←
wood	350kg stere	1200←	1200/stere	1.00	30←	3.33←

Table 9: Economics of Transport: Wood vs. Charcoal

km of haul	haul + production cost per 10,000,000 Kcal	
	WOOD	CHARCOAL
85	33	34
95 ←	36 ←	36 ←
100	38	37

Table 10: Economics of Transport: Wood vs. Charcoal

Assumptions	km of haul	haul + production cost per 10,000,000 Kcal	
Open fire efficiency 7% Charcoal Stove efficiency 25%		WOOD	CHARCOAL
	5 ←	71 ←	88 ←
	10	100	92
	15	128	92

Table 11: Economics of Transport: Wood vs. Charcoal

Assumptions	km of haul	haul + production cost per 10,000,000 Kcal	
Wood Stove efficiency 30% Charcoal Stove efficiency 25%		WOOD	CHARCOAL
	140	170	172
	145	176 ←	176 ←
	150	183	180

Table 12: Economics of Transport: Wood vs. Charcoal

Assumptions	km of haul	haul + production cost per 10,000,000 Kcal	
Wood Stove efficiency 30%		WOOD	CHARCOAL
	50	70	72
Charcoal Stove efficiency 40%	55 ←	73 ←	75 ←
	60	80	77

Tables 5 – 8 show that charcoal will be the cheaper fuel for urban use, unless an improved wood stove program is initiated in the cities. But if wood is used then there will be air quality problems.

Tables 5 – 8 don't take into account the transportation advantage of charcoal. Let us look at our previous transportation example from page 8. The break even point in this case is at 95 km (Table 9). This assumes that all of the energy value that is hauled will be used; however, 100% efficient stoves are not possible.

Tables 10, 11, 12 show the effects of different stove efficiencies on comparative transportation costs between wood and charcoal. If we assign the values of 25% efficiency for charcoal stoves and 7% efficiency for open wood fires the break even point is between 5 and 10 km (Table 10).

If we start a program to improve wood stoves and succeed in raising their efficiency to 30% the break even point is then about 145 km (Table 11). If we also work on charcoal stoves and improve their efficiency to 40%, the break even point will be between 55 and 60 km (Table 12).

In summary, charcoal can fill a role as an economically reasonable cooking fuel even in an area of fuel wood shortages. It is especially useful in meeting requirements for an urban fuel.

2. Charcoal and Forestry

Wood is a regenerative source of energy, but in nearly all developing countries wood reserves are decreasing. Responsible for the rapid deforestation, however, are not the small charcoal makers and the wood-collecting housewives, but extensive shifting cultivation and increasing demand for agricultural land. It will be the most important task for national forestry administrations to keep the exploitation of forests under control.

However, if a forest is not simply cleared to get valuable logwood, but cultivated to get a sustained yield, a considerable amount of wood can be extracted every year by thinnings, stand improvement practices and wood residues after logging. Here are potential sources of income for wood collectors, charcoal makers and haulage contractors to produce and distribute the charcoal for the urban population.

In areas with thin secondary forests or already degraded vegetation the making of charcoal for the requirements of cities would be irresponsible. For the rural population of such areas the fuel-saving cookstove with wood as fuel is the best solution (see: "Fuel-Saving Cookstoves", Aprovecho Institute 1984, available from GATE/GTZ).

Reforestation is the task of the future to meet the fuelwood needs of the population and to save the environment. But a restraint on such a "social forestry" is funding. Providing for indigenous needs doesn't bring in foreign capital. Money spent on extension services to teach people to provide for their own needs brings in nothing and has fairly high costs when considered over the short run. It is only when such policies are considered over the long run in the context of national economic health that the benefits become apparent. For example, a program to encourage households to plant their own wood lots can have future returns in fuelwood and building materials for the family; and in seedlings, fuel, poles, and building materials for local and regional markets. Trees protect soil from erosion, raise water tables, and protect homes from wind and dust. The dependence on the government to provide cooking fuel is removed. This may result in the country becoming less dependent on imported fossil fuels.

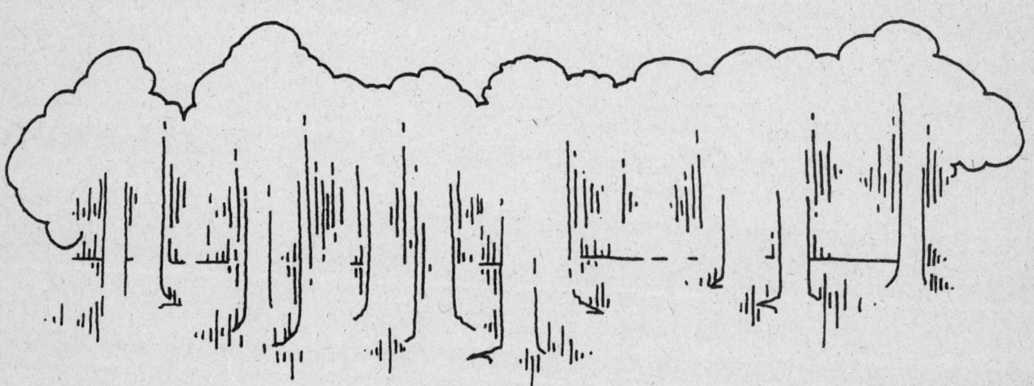

3. Working with Charcoal Makers

A project in charcoal production and use can have an advantage for the development worker over a project in improved wood-using cookstoves. Extension work involving wood-using cookstoves requires working with an entire population. Since charcoal makers and charcoal stove builders are a very small percentage of the population, by working with this small group of people you can affect all charcoal users. This greatly increases the speed at which a program can be introduced.

Necessity of Working with Charcoal Makers

Before getting involved in charcoal production it is important for you to understand where the expertise lies. This book contains a mere fraction of the information available on charcoal. The bulk of the information is not to be found in books or universities; rather, it resides with the smallscale indigenous charcoal makers themselves. Through long years of trial and error, they have learned a great deal about what works and what doesn't. Their knowledge is particularly apparent in adaptation of kiln designs to local conditions and in kiln operations.

This situation places the development worker much more in the position of a student than in the role of a teacher or expert. It will be the development worker's task to learn from the charcoal makers and, together with them, improve charcoal production.

Who Charcoal Makers Are

Because charcoal makers are such a diverse group there is no single description of them. However, the following points touch on the ways in which charcoal makers can vary and will give you some guidelines of what to look for when you define the group "charcoal makers" in your area.

For some charcoal makers, charcoal making may be the only source of income in contrast to parttime charcoal making for cash income along with subsistence farming. Charcoal makers may be from one or more local ethnic groups or they may be a social or ethnic group which comes from outside the geographic area. They may even be semi-nomadic, living near their work and constantly moving from kiln to kiln. They may be semi-nomadic for part of the year and farm for part of the year. Women may or may not participate directly, or may participate but only in certain phases of production. In some cases women actually do the charcoal making in conjunction with the clearing of dead material around their food farms. Charcoal making is often a traditional occupation, and knowledge may be passed from one generation to another. In other cases, however, knowledge is passed on through a master-apprentice relationship. The charcoal makers may be organized collectively in a cooperative, work in a near-serf relationship with a patron, or there may be no unifying organization above the level of the family unit.

Charcoal making may be restricted to a specific ethnic or social group and there may be some type of social status attached. In certain cultures the development worker's effectiveness may be hindered by the status of the charcoal makers. He may suddenly find himself placed in a social position such that he can work with charcoal makers and no one else. This is one reason why charcoal making may not work as a secondary undertaking for the development worker.

Finding Charcoal Makers

To find charcoal makers just ask people where charcoal is made and they will usually direct you to the general area. After you find the general area of production, look for bags or piles of charcoal along the road. The production site will be nearby. This sometimes won't work if you ask the retail charcoal venders, since they may think you are going to buy directly from the charcoal maker. However, if the charcoal makers transport and market their own production, as is sometimes the case, then sooner or later all of the charcoal makers in the area will pass through the market.

Getting Information from Charcoal Makers

One of the best methods of getting local charcoal information is just to show an interest. Charcoal makers are usually quite flattered that you are interested in their work and will teach you all they know. In this situation you are a student and should act accordingly. Once you are accepted by the charcoal makers, pick up as much information as you can. This is not the time to try to introduce change.

There is a certain crafts brotherhood feeling among charcoal makers that cuts across cultural boundaries. In one case, a researcher arrived at a charcoal site and, after the proper greetings were exchanged, announced that he, too, was a charcoal maker. The local charcoal maker immediately grabbed the hand of the stranger and inspected it for calluses. After satisfying himself that this stranger was a co-worker he even allowed the researcher to take photographs, a thing not normally permitted in this particular culture. If you intend to claim brotherhood privileges, be prepared to be tested as pretension is easily seen through.

Extension Work

After working with charcoal makers in a successful kiln improvement research program which includes yield tests, trial runs, and user acceptability tests, you will be ready to start extension work.

Although a patient and well trained outsider can be an effective extension worker, trained nationals are often the best extension agents since they already know the language and customs of the locals. However, there are some situations where the outsider has more status and can be more influential.

It is best to exchange techniques by demonstration and participation, allowing the charcoal makers to participate and experience as much as possible. A working model of your kiln is worth much more than many hours of lecture. Charcoal makers are curious about other methods of charcoal production, and a running kiln can elicit many criticisms, comments, and occasionally even praise. Be open to their suggestions for possible improvements that can be made. For example, the air chamber that makes the Casamance kiln (page 27) so easy to run came from the traditional charcoal makers who were working on the project. They couldn't explain why the apron (the outer wall of the circumferential air chamber) should be placed on the ground and not on the platform, and the "expert" couldn't talk them out of it, so it was tried and found to be the most effective ingredient in the design.

When demonstrating your kiln designs, emphasize all the advantages such as efficiency, safety, and ease of operation. You will probably find that the three most effective selling points are:

1. the kiln takes less time to carbonize;
2. it takes less work;
3. it makes more charcoal (money).

You have to keep in mind, however, that "less time to carbonize" may result in insufficient carbonization, and an increase of charcoal yield (measured in kg charcoal per kg wood) can also mean a high content of uncompletely carbonized wood in the charcoal. So be careful: you always have to produce good quality charcoal which is completely degassed of tars and visible smoke.

The use of a demonstration site for extension work can be a help or a hindrance. It does give a comfortable place to do your research, some amount of implied status, and easy access to tools, but it may be difficult to get busy people to come and see what you are doing. It also roots you to one spot. It may even tend to increase the cultural and economic distance between you and the charcoal makers.

4. Carbonization

Charcoal is wood charred in the absence of air. In slightly more technical terms, the actual process is destructive distillation of organic matter in a near oxygen-free environment. During this process, water, tars and other substances are driven off as heat is applied. What is produced is a substance composed mainly of carbon (80% +) but which may also include hydrocarbons (10–20%), ash (0.5–10%), and traces of various minerals such as sulphur and phosphorus. Due to its lack of moisture and high carbon content, charcoal contains large amounts of energy. It has twice the amount of energy as an equal amount of air-dry wood.

The process by which organic material becomes charcoal is called carbonization, and can be broken into the four distinct phases of combustion, dehydration, exothermic, and cooling. All of these phases may be going on in the kiln at the same time, but each piece of wood must pass through this four-phase sequence. The time required for each of the different stages is dependent on size of kiln, type of kiln, operations procedure, moisture content of the wood, and weather conditions. These will be discussed in chapter 7.

Combustion

The combustion phase is the only time during the carbonization process when large quantities of oxygen are required. In this phase a fire is started in one part of the charge until that section is burning well. During this stage the kiln is heated from ambient temperature to over 500 °C (Fig. 4-1). After combustion is completed the oxygen is drastically reduced and the temperature of the kiln drops to as low as 120 °C (Fig. 4-2).

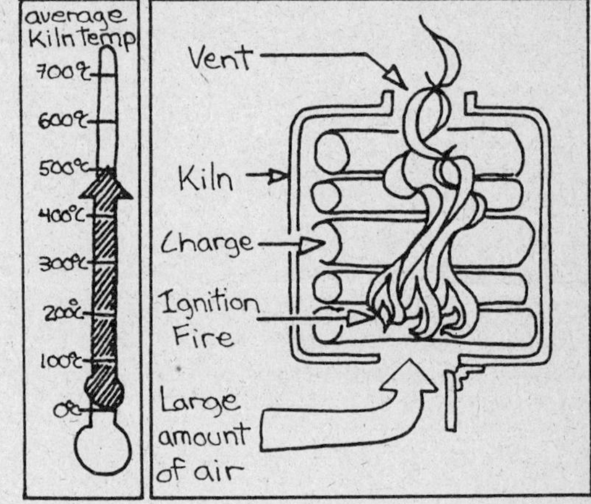

Fig. 4-1: Combustion

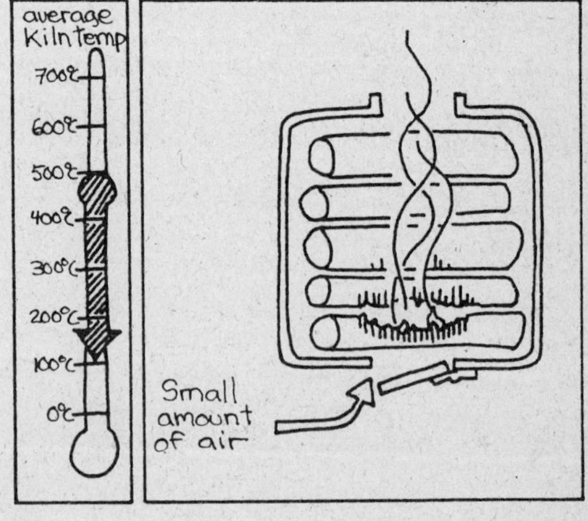

Fig. 4-2: Reducing the Oxygen

Dehydration

The heat provided by combustion drives moisture out of the charge in the form of steam. As the charge dries, the temperature rises slowly to about 270 °C. Dehydration continues until all of the free moisture is driven off. The steam is white, thick and moist (Fig. 4-3).

Fig. 4-3: Dehydration

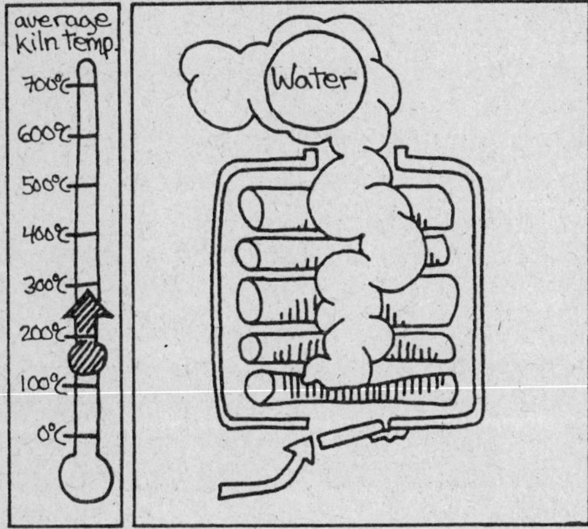

Exothermic

Once dehydration is completed, the wood itself starts to break down by the influence of heat, producing even more heat by exothermic reactions. The air is completely cut off at this point to prevent combustion. The main gaseous products of this thermal decomposition (pyrolisis) are acetic acid, methyl alcohol and tar. The remaining solid material is the charcoal.
During this exothermic phase the temperature rises to 700 °C. The smoke is yellow, hot and oily (Fig. 4-4).

Fig. 4-4: Exothermic

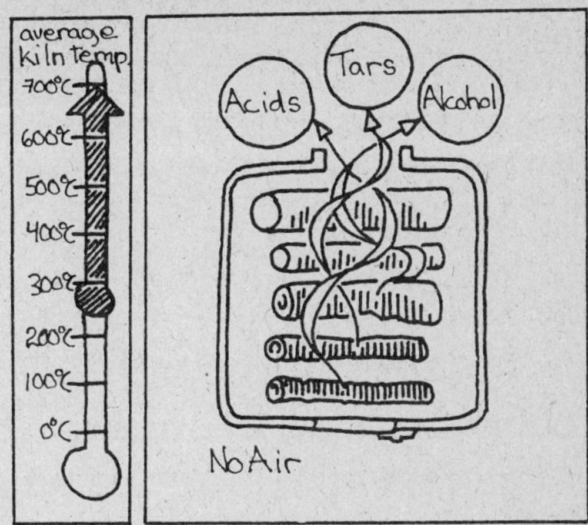

Cooling

The cooling stage in the kiln allows the temperature of the charcoal to drop to a point where it can be extracted from the kiln (Fig. 4-5).

Fig. 4-5: Cooling

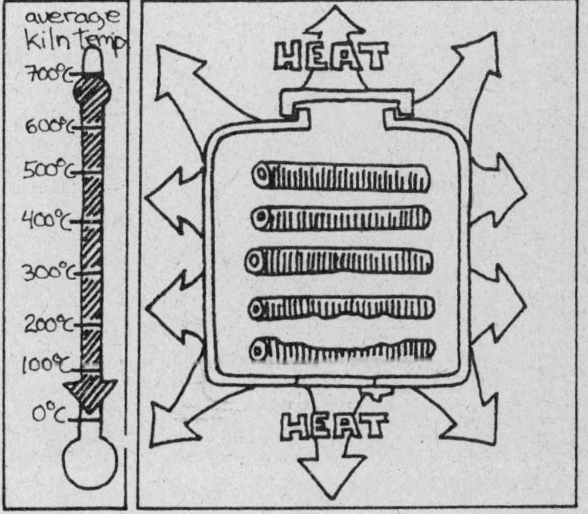

5. Charcoal Making Devices

The function of charcoal making devices is to control the amount of oxygen which can get to the wood. This is accomplished with a nonflammable cover. Charcoal making devices can be classified either as *kilns* (in which a portion of the charge is burned to start the carbonization process, Fig. 5-1) or as *retorts* (in which the heat needed to carbonize the charge is applied from outside the shell of the device, Fig. 5-2). Retorts will be discussed later in chapter 8.

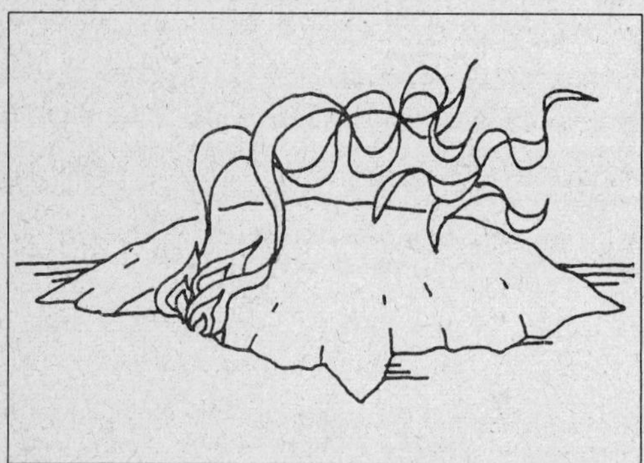

Fig. 5-1: Kiln

Fig. 5-2: Retort

Essential Parts of a Kiln

In order to function kilns must have the following:

1. *Charge of organic matter.* Any organic material from grass to trees can be made into charcoal. The kind of raw material you start with will determine the type of charcoal produced and the ease of production. Hardwoods will produce dense, hard charcoal, while softwoods will produce light, softer charcoal. Charcoal made from bark will have a higher ash content which can cause problems in some industrial applications. Small, finely divided material such as grass or sawdust will produce charcoal which will require briquetting for normal use. Resinous wood may yield a charcoal that produces a flame from residual resins and may impart an unpleasant flavor to food.

2. *Heat source.* Some provision must be made to light the kiln. Ignition can either be through lighting part of the charge or by dumping hot coals from an external fire into the kiln.

3. *Cover.* The cover controls the amount of oxygen which reaches the charge. It can be made from a variety of different materials including earth, metal, masonry, cement and adobe (Fig. 5-3).

4. *Vent.* Small amounts of air are required to sustain the carbonization process until the exothermic reaction takes over. This function is performed by the vent, which may be

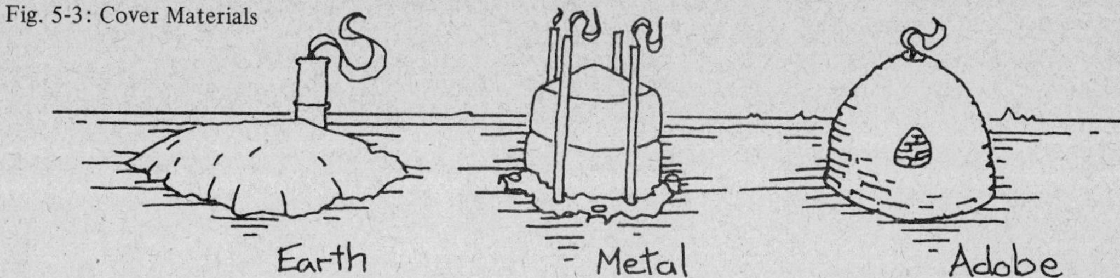
Fig. 5-3: Cover Materials

a hole poked through a cover, a sandy earth cover which allows in a certain amount of air, or even elaborate automatic controls.

5. *Exhaust.* In some cases the intake and exhaust functions may be carried out by the same opening. Usually there is a separate exhaust port, which may consist of only a hole poked through the cover or may be a separate structure such as a chimney.

Characteristics of a Good Kiln

A good kiln is one that can satisfy a wide range of objectives while still producing high yields. Let's take a look at some of these characteristics:

Socio-cultural Fit

The kiln must fit local conditions and correspond to what people need, to what they can afford, and to their skills. The kiln should be developed with the people who will eventually use it. In this way the kiln becomes *their* technology and is much more likely to be accepted.

Low Cost

Common sense dictates that a tool should be as low in cost as possible while still accomplishing the job as intended. Using low cost, locally available materials should be empha-

sized. Another thing to keep in mind is the lack of capital among most charcoal makers. There are not many charcoal makers who can efford even a relatively modest investment with the hope of improved efficiency paying off the costs involved.

Long Gas Path

A long gas path ensures that the heat produced by carbonization is available to heat up the rest of the charge. In a kiln there are two ways of directing hot gases: direct draft and reverse draft. In a direct draft kiln (Fig. 5-4) the air enters at the bottom, feeds the carbonization process, and exits at the

Fig. 5-4: Direct Draft

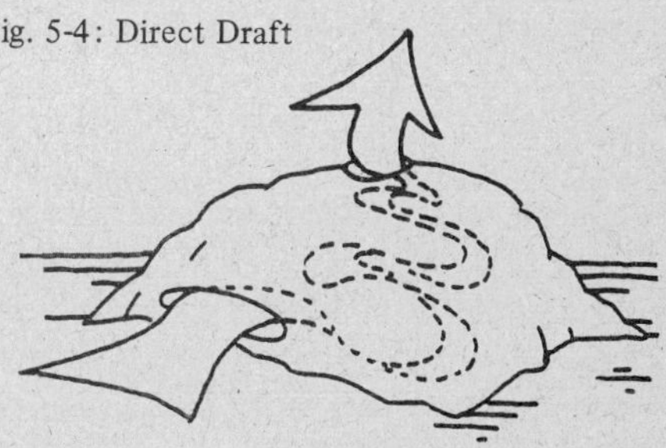

Fig. 5-5: Reverse Draft

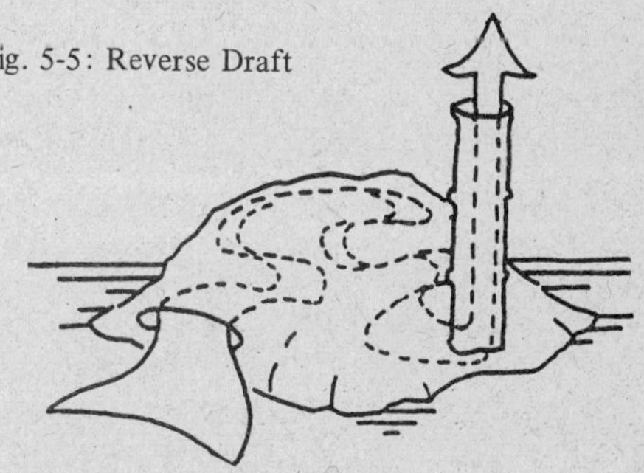

top of the kiln. A reverse draft kiln (Fig. 5-5) is started in the same manner as a direct draft, but after heating up, the gases are circulated first up and then down. The gases pass of heat, pick up moisture from the charge, and finally exit from the bottom of the kiln into an external chimney which is necessary to provide an up draft.

Tight Packing

Tight packing (Fig. 5-6) allows more wood to be packed into a kiln without increasing the kiln size. It also makes for a very long gas path as the gases wind their way through the kiln. Additionally, tight packing lessens the occurrence of cave-ins in earth covered kilns, since the even density of the charge allows it to settle evenly as it shrinks. Heat transfer within the kilns is also enhanced by tight packing. The packing should break up any favored air channels since the wood along a favored air channel may burn completely leaving the rest of the charge virtually untouched (Fig. 5-7).

Proper Size

In general, a larger kiln will be more efficient. There is, of course, a limit. Unless heavy equipment is being used, the kiln must remain a size that humans can work with. Considerations such as the amount of labor available, the amount of materials available, and how far earth can be thrown to cover a kiln must be taken into account (Fig. 5-8).

Volatiles Collection

Volatiles can be condensed by using baffled chimneys. Condensed volatiles may be used for wood preservatives or burnable fuels or further refined into basic chemicals.

Fig. 5-6: Tight Packing

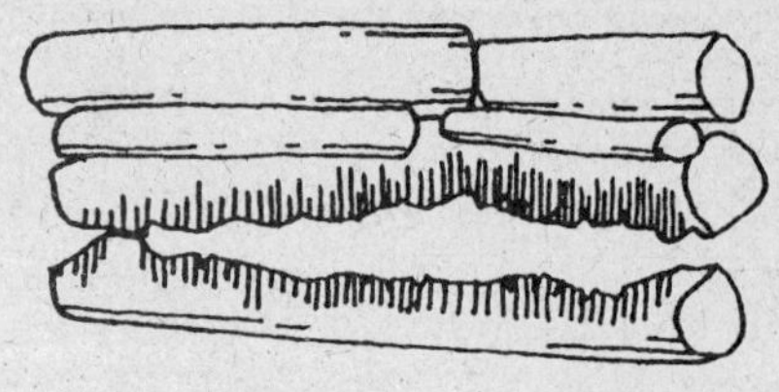

Fig. 5-7: Results of a Favored Smoke Path

Fig. 5-8

Fig. 5-9: How Condensation Works

No Hot Spots

The proper amount of air should be evenly distributed to promote even carbonization. One way of accomplishing this is to place the charge on an elevated platform to assist in getting the air where it is needed (Fig. 5-10). The Casamance kiln takes this one step further through the use of a circumferential air exchange chamber around the base of the kiln.

Speed

The faster a kiln will carbonize without a reduction of quality or quantity, the more charcoal can be made and the greater the income for the charcoal maker.

Fig. 5-10

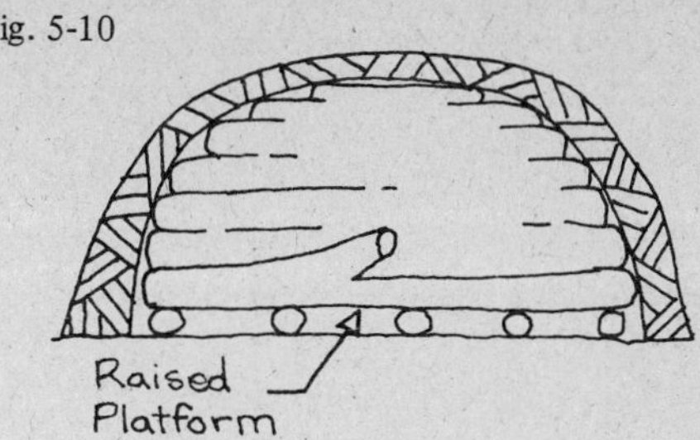

Efficient Use of Labor

Notice that this doesn't say "less labor." It isn't economic good sense to substitute capital for labor in a labor surplus economy. This is especially true in rural areas where lack of cash income opportunity is the basis of urban migration problems.

6. Kiln Review

In this section we will take a look at a wide variety of kilns, discuss their merits and weaknesses, and give a basic idea of how they are constructed. None of these kilns may match your particular needs, so be aware that you may have to modify one or develop your own.

Earth Covered Kilns

The simplest type of kiln is a neatly stacked pile of wood which is then completely covered with green vegetation and earth (Fig. 6-1). Almost any size of wood can be carbonized in earth covered kilns as they can easily be modified in shape and size to suit your needs. In round kilns wood is stacked radially; however, in a simple rectangular kiln the wood can be stacked either lengthways or crossways.

Although the method of cutting and stacking in Fig. 6-2 takes more time, it results in a better operating kiln and a higher yield than the kiln in Fig. 6-3 which is easy to construct but may burn out along the air channels.

Radial stacking with a central lighting point (Fig. 6-4) takes more skill to construct but provides for more even carbonization.

Larger pieces are stacked in the center. This is a general rule which applies to most kilns.

Enough green, slowburning vegetation is applied to completely cover the charge. This keeps the earth from falling into the kiln where it can cause incomplete carbonization. Usually two or three layers of vegetation are sufficient to do the job. A layer of earth about 25 cm thick is then placed on top of the vegetation to act as an air seal. If the soil is sandy, this layer may have to be thicker (Fig. 6-5).

Holes are poked through the vegetation and earth cover to act as air vents. This kiln is

Fig 6-1: Earth Covered Kiln

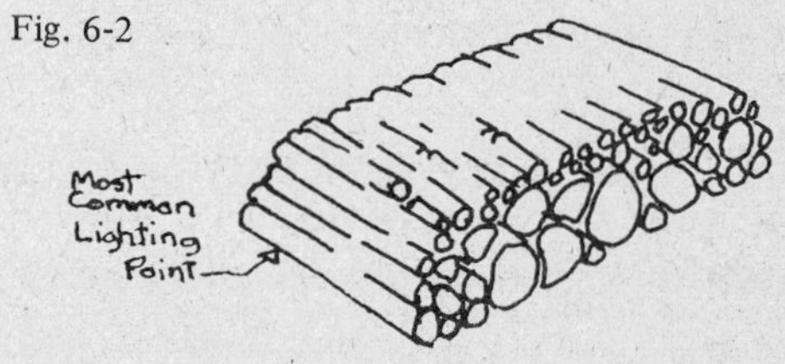

Fig. 6-2

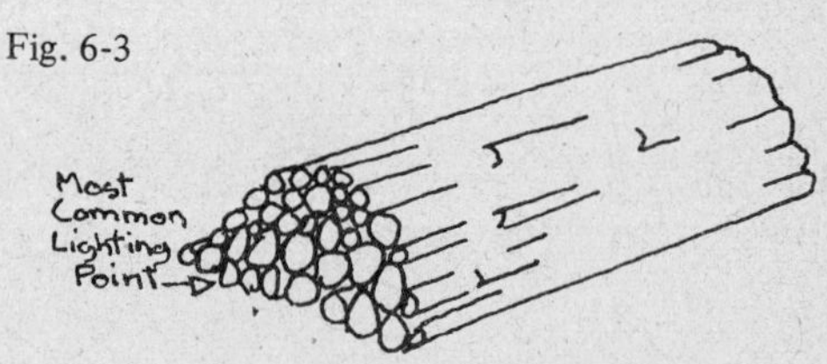

Fig. 6-3

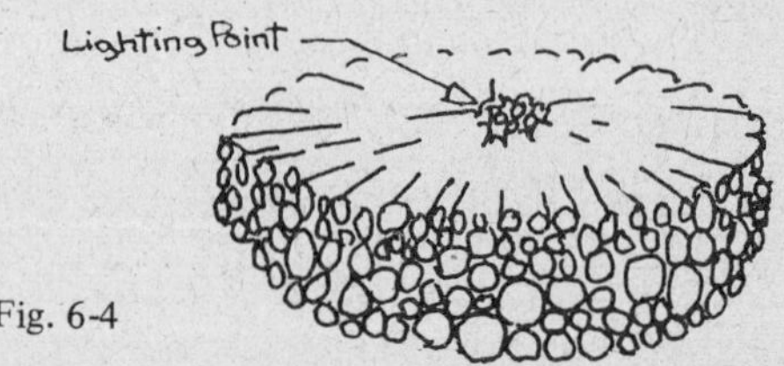

Fig. 6-4

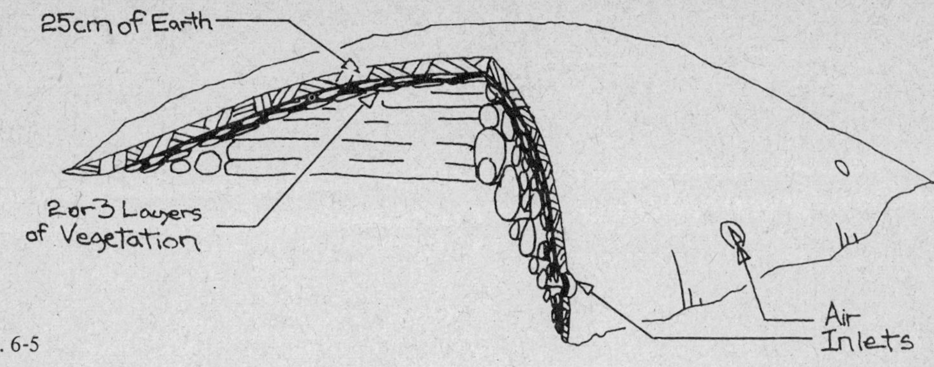

Fig. 6-5

usually lit on one end and is operated with direct draft. The main drawback with a kiln this simple is that the gas and heat transfer are not even, which can cause much of the charge to either burn up completely or not to carbonize at all.

The following applies to all earth covered kilns:

Advantages
— No capital required.
— Uses local materials.
— Traditional skills.
— Easy to understand.
— Different phases of carbonization are easily observed.

Disadvantages
— Must be rebuilt with each carbonziation cycle.
— Prone to collapse.
— Contaminates the charcoal with earth.

After the carbonization run, the earth that is used as a kiln cover is full of small charcoal particles called fines. Fines can be recovered by dumping the earth in vats of water and scooping off the fines when they float to the top. After drying they can be made into briquettes.

Woodwall Kilns

A more sophisticated type of kiln uses wood planks or poles to hold the soil against the side of the charge (Fig. 6-6). This allows a smaller amount of soil to be held in exactly the right place, which saves a lot of work when compared with the other types of earth kiln. This kiln employs palm fronds as the vegetative cover for the sides. These fronds also poke up through the top and act as air vents and chimneys. The charge is supported on stringers (described below) which gives gas circulation space beneath the charge and makes for a more even carbonization. One disadvantage of this type of kiln is that wood must be cut and stacked very accurately with the sides flush to prevent wood from poking through the cover.

This kiln usually is 1 to 2 meters high, 1 to 2 meters wide, and 3 meters long. Two poles 10 cm in diameter by 3 meters long are laid parallel on the ground as stringers. The rest of the charge is laid across the stringers.

Once the desired height is reached, stakes about 2 meters long are hammered into the ground about 30 cm from the sides and ends of the charge. Three stakes at each side and two at each end are usually sufficient. Palm fronds are cut and stuck, small end down, into the ground next to the

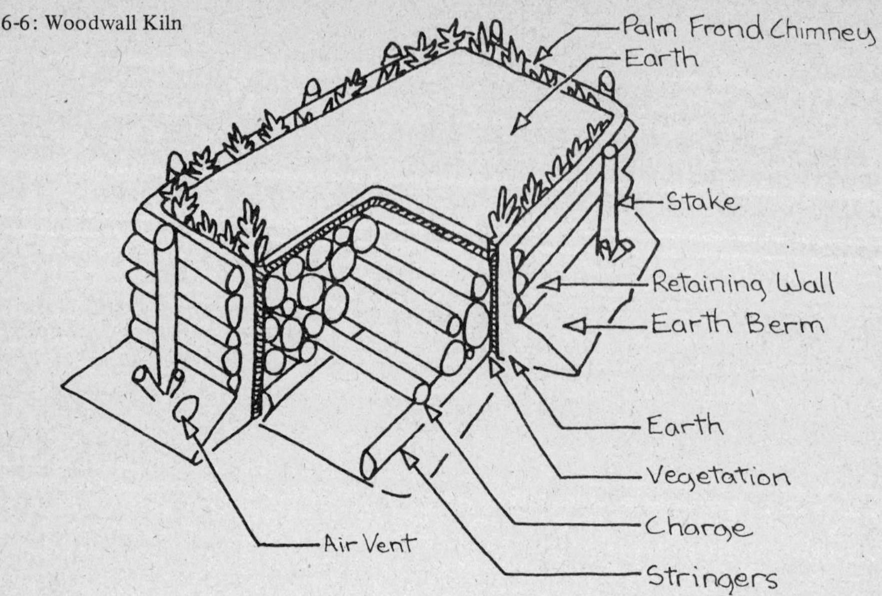

charge. The tops of the fronds should extend about 50 cm above the charge. Earth is then piled against the sides until it naturally inclines beyond the stakes. Planks are then placed behind the stakes and more earth is added. The earth should be tamped to make a better air seal. This process continues until all the sides except the one to be lit are covered. The top is covered with vegetation and soil. The kiln is now ready to be lit.

Fig. 6-7: Casamance Kiln (adapted from a drawing by Kevin Stewart, P.C.V.)

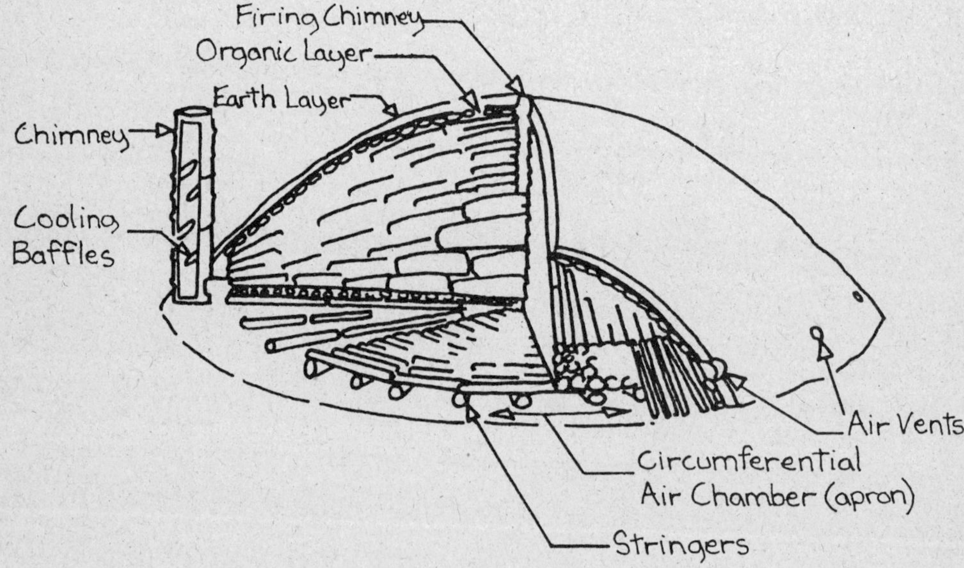

Casamance Kiln

The Casamance Kiln (Fig. 6-7) is an example of an improved traditional kiln. In the original the charge was arranged radially, directly on the ground. This structure was then covered with vegetation and finally with earth. Ignition was started down a central lighting chimney. The kiln was operated by direct draft, with intakes at the base and exhaust ports about 3/4 of the way up the sides.

The first improvements were to build a platform to raise the charge up off the ground and a circumferential air chamber that extended all around the kiln. Next a single large chimney was added on one side to reverse the draft. This then became a very sophisticated gas and heat circulation system which remained simple to construct. It resulted in dramatic yield increases and shortened carbonization time for very little capital.

Construction of the Casamance Kiln is commenced by laying out stringers in a radiating pattern. On these a platform of poles is laid. Large pieces are stacked in the center and spaces are filled with smaller material. In the center of the top a one meter deep hole is left for ignition. The circumferential air chamber is constructed by leaning sticks against the side of the stack.

The chimney is made of three 220 liter drums welded together. Baffles within the chimney serve to increase the condensation surface for volatile collection. The chimney is placed inside the air chamber and draws gases from under the platform.

The kiln is covered with vegetation and earth. It is usually ignited with coals placed in the ignition hole in the top.

Pit Kilns

The most universal type of kiln is the pit kiln. In its simplest form it is a hole in the ground in which a fire is started, the wood added, and the top covered with vegetation and earth (Fig. 6-9). Sometimes a metal cover is used for a better air seal and to reduce contamination.

Pit kilns can be improved by using stringers under the charge and chimneys with intakes located low in the corners of the pit. This greatly improves gas distribution within the kiln and allows it to be run as a reversed draft kiln. To avoid the added cost of chimneys, holes can be bored in the ground (Fig. 6-8).

Advantages
- No capital needed.
- Good for repeated use.
- Accepted technology.

Fig. 6-8: Improved Pit Kiln

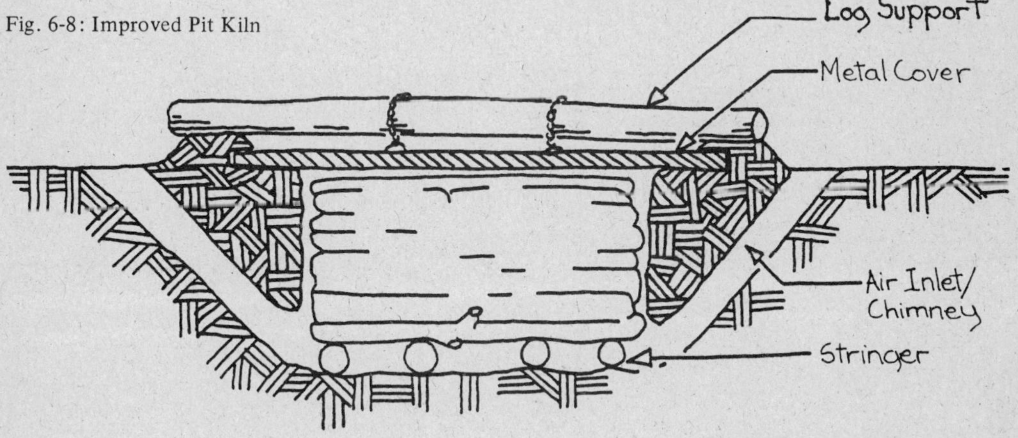

Disadvantages
- Hard to dig.
- May flood during rains.
- Hard to load and unload.
- Heat is lost to the ground around the kiln.

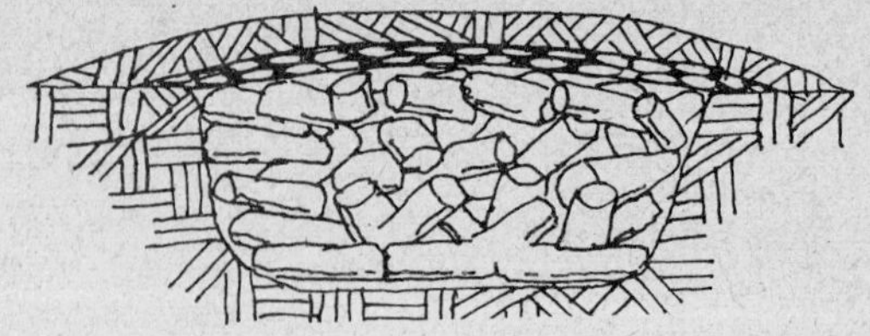
Fig. 6-9: Pit Kiln

Metal Kilns

Metal is often used in kilns either alone or in combination with other materials.

Advantages
- Quick cooling.
- Airtight.
- Won't collapse.
- Portable.
- Strong.

Disadvantages
- Can warp or burn out under heat.
- Expensive.

- Usually need welding equipment for construction and repair.
- Radiates heat.
- Heavy.

Mark V

The most widely known type of metal kiln is the Mark V (Fig. 6-11). It has a main body of two cylinders joined with a slightly conical lid on top. The lid has a hole in the center which is capped except during ignition. Joints between the three main parts are sealed with sand.

Fig. 6-10: Metal Kiln

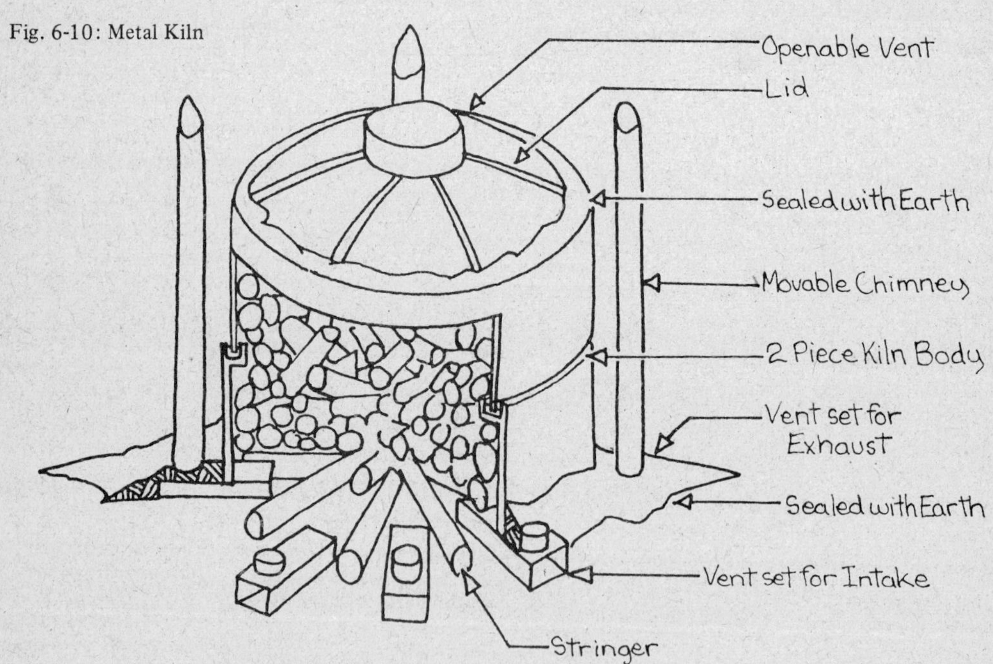

28

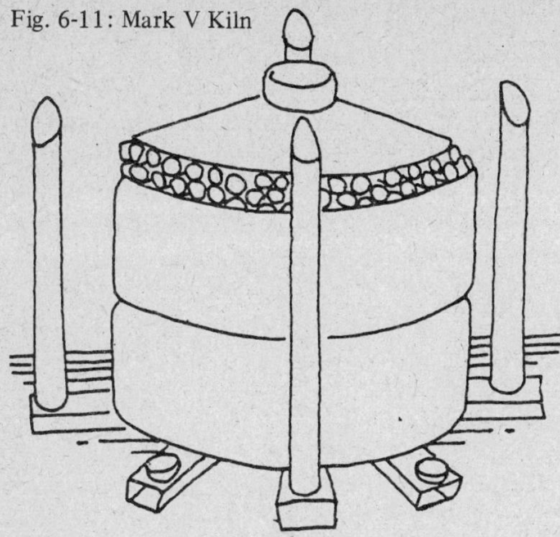
Fig. 6-11: Mark V Kiln

This kiln operates by reverse draft. There are eight vent channels spaced evenly at the bottom. These alternately function as air intakes, or, with chimneys added, smoke exhausts.

This type of kiln is expensive as it is made of heat resistant steel. If treated with care it will last three to five years, but if the operator is careless it can be useless after the first firing. Cheaper models made locally from mild steel will degrade even faster.

To set up the kiln, lay the eight vents radially and place the bottom ring on top. Lay stringers between vents to prevent direct gas passage between the vents.

Wood is then stacked in the kiln until the bottom ring is full. The top ring is placed and further wood added until it reaches about 10 cm above the top ring. The lid is then positioned on top of the charge so that it will settle onto the top ring when ignition takes place.

The chimneys are placed on every other vent. The kiln is lit by dumping hot coals in the top vent. Lighting from the top will keep the intense heat in the center of the charge and away from the kiln sides. After the top has settled into place seal the top cap and check all of the sand seals. Seal the gap between the vents at the base with earth, leaving only the vents open.

Adobe and Masonry Kilns

Adobe is a traditional building material in Central and South America while masonry block has been used for kilns in North America.

Advantages
— Easy to use.
— Insulating properties.
— Masonry: can be disassembled and moved.
— Adobe: No capital investment.

Disadvantages
— Not portable.
— May crack.
— Adobe: brick must be made.
— Masonry: expensive, needs special aggregate.

Beehive Kiln

The beehive (Fig. 6-12) is an adobe kiln common in South America. It is constructed of brick with straight walls and a domed roof. An important feature is the steel reinforcing ring at the base of the dome. This constricts the lateral force exerted by the dome and keeps the wall beneath from being forced outward. In construction the walls are laid dry. The outside is initially plastered with lime plaster as a gas barrier and patched before each use. After loading, the door can simply be laid up with brick and plastered. This probably provides a better seal than trying to fit a metal door and is certainly cheaper.

There are various vent arrangements around the base or up the sides. There is usually a hole in the top of the dome which is closed after ignition, thereby establishing a reverse draft exhausting through four external chimneys.

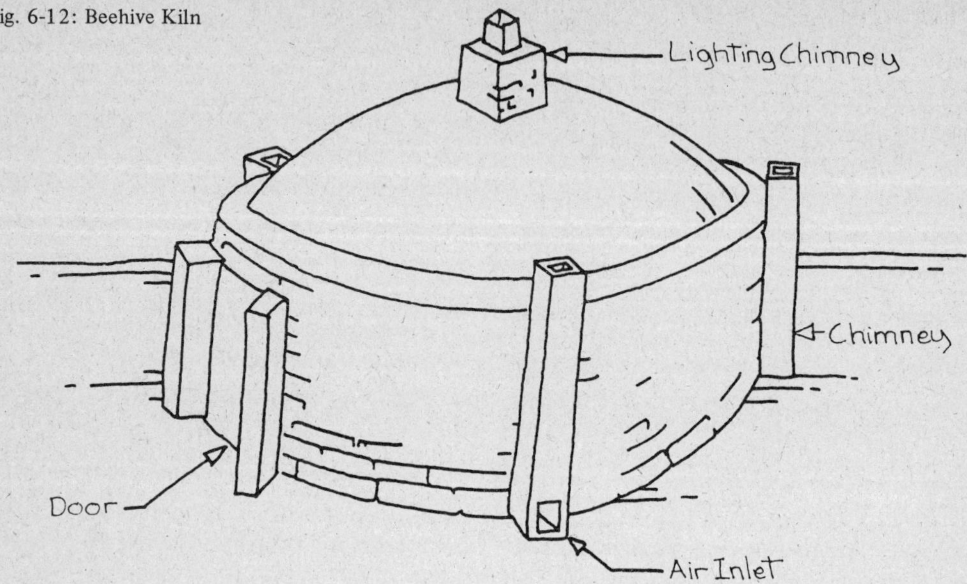

Fig. 6-12: Beehive Kiln

Cement Kilns

Cement kilns used to be in common use in the United States and a few have been built in developing countries.

Advantages
— Permanence.
— Durable, if built well.

Disadvantages
— Expensive.
— Not portable.
— Requires special aggregate.

Missouri Kiln

The best known cement kiln is the Missouri Kiln (Fig. 6-14). This mammoth kiln takes up to one and a half months to complete a carbonization cycle. It has alternate chimneys and vents along the sides, and four lighting vents in the top. There are large metal doors in each end to facilitate mechanized loading and unloading. The expanded shale aggregate needed for construction is not widely available and use of common sand results in a mix that cracks under heat.

Mixed Kilns

These are combinations of pit kilns and portable metal kilns. The pit is much shallower than in regular pit kilns. This makes it easier to dig, load, and unload. Metal kilns usually start degrading at the base where the air inlets are and where the kiln is the hottest. In mixed kilns this hot area is underground which can mean a much longer life for the kiln (Fig. 6-15). This also slows heat loss from radiation during the early stages of carbonization. Sometimes metal, adobe, and pit kilns are combined, and occasionally pit kilns are covered with earth (Fig. 6-13).

Fig. 6-13

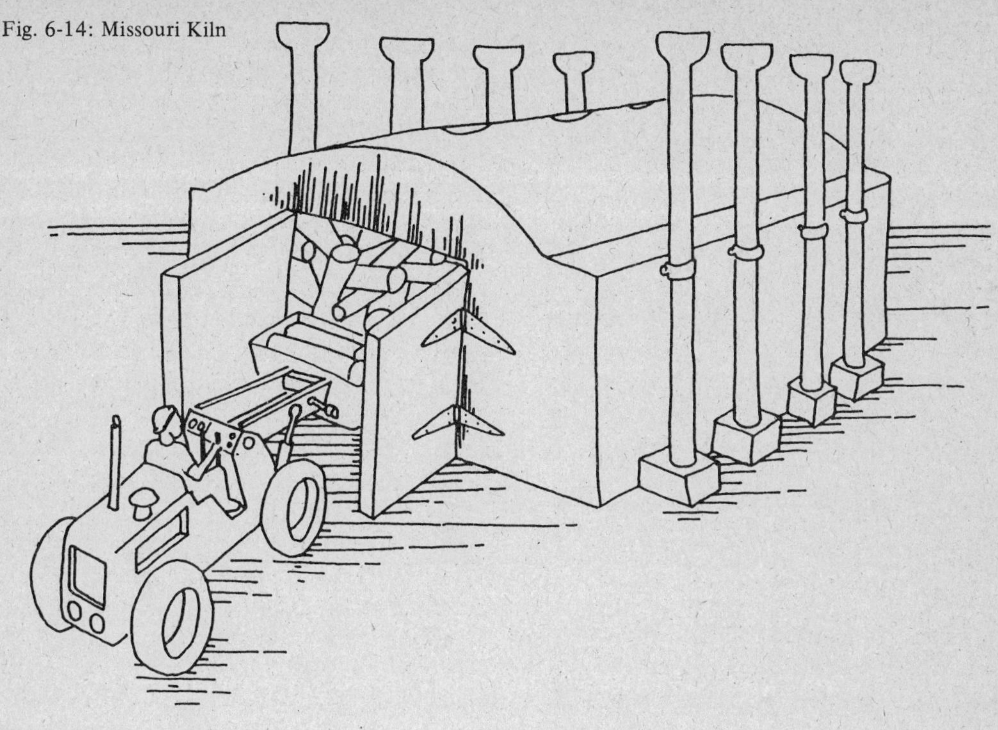

Fig. 6-14: Missouri Kiln

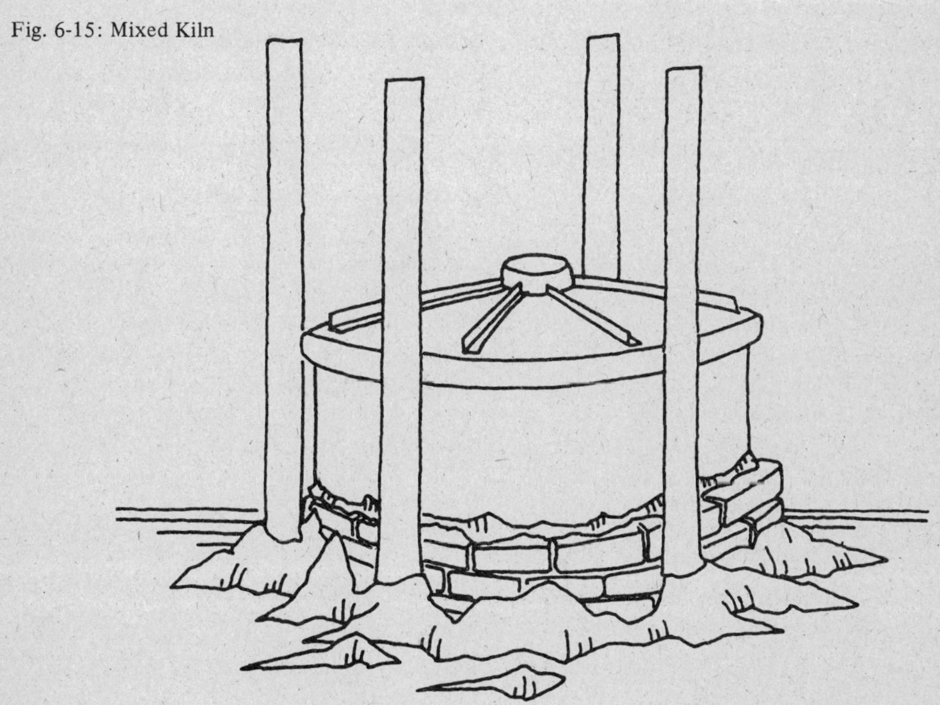

Fig. 6-15: Mixed Kiln

7. Kiln Operation

The operation procedure outlined in this chapter applies to most kilns. Only basic hand tools such as shovels, axes, saws, and rakes are needed for kiln operation. Charcoal making devices that differ in operation will be covered in chapter 8.

Weather Influence

Weather conditions will influence kiln operation. The most obvious is rain, which increases carbonization time and lowers yield somewhat. Soil moisture can also determine how easy an earth kiln is to construct and extinguish.

Site Selection

Select a kiln site close to the wood source. Clear rubble and loose earth from an area that will leave three meters around the completed kiln. This is done not so much for fire safety but to give a clean surface to spread the charcoal and decrease the amount of impurities mixed in with the charcoal.

Wood Gathering

The next step is to cut the wood. Normal is a length of one meter and a maximum diameter of 20 cm. Some kilns require uniformity in wood size, species, and moisture content, while others are more tolerant of variation. It is nice to have some small material available for tight packing.

Kiln Preparation

The kiln is set up or constructed on the site and charged with wood. The method of wood placement will differ for each kiln type (see chapter 6).

Lighting the Kiln

A good bed of coals is needed in the charge in order to start the carbonization process. There are two ways of doing this.
The first is to start a fire in a small section of the kiln (Fig. 7-2). This method takes a certain amount of skill to get sufficient coals without burning too much of the charge. The second is to place hot coals from a separate fire into the charge (Fig. 7-3). This method is easier to control and can raise the temperature faster.
Ten to twenty minutes after lighting, thick smoke should be billowing out of the kiln. If the wood is very dry this time will be shorter. At this point any ignition ports should be closed but air vents should remain

Fig. 7-1

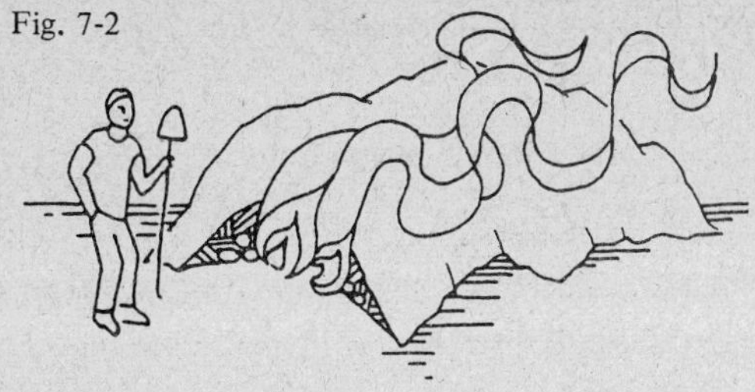

Fig. 7-2

open. If it is a reverse draft kiln, the chimney(s) should be open when the top vent is closed.

Once the ignition ports are closed the volume of smoke will reduce drastically. The kiln may look dead but it is adjusting to the lowering of oxygen levels and the loss of heat to the rest of the charge. In time it should start to pick up.

Fig. 7-3

Operating the Vents

By manipulating the vents you are providing just the right amount of air to keep the carbonization process going, while not supplying so much air that the charge is in danger of burning completely. Watch the kiln carefully: different wood types, stacking methods, and weather conditions can affect how the vents should be set. These will have to be learned by experience as the effect will vary in each different kiln type.

In some respects the vents will operate automatically, changing from inlet to exhaust depending on the needs. Even a chimney will, at times, act as an inlet. A vent should be closed when the carbonization front has passed, or when glowing coals can be observed through the hole. In metal kilns such as a Mark V, the chimneys and vents are switched halfway through the run to promote more even carbonization (Fig. 7-4).

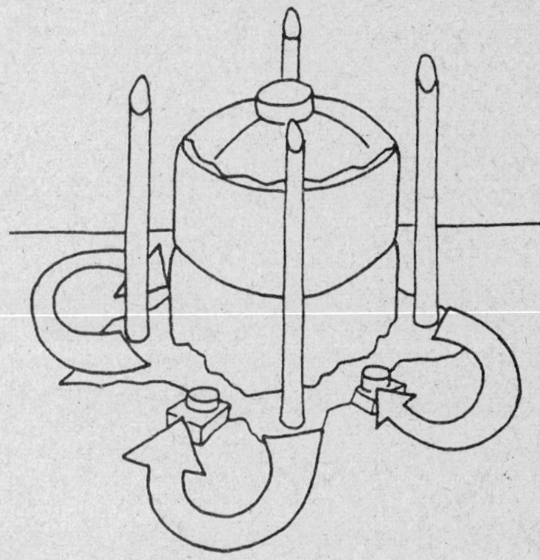

Fig. 7-4: Operating Vents in a Mark V Kiln

Fig. 7-5 is an example of how to operate the vents in a woodwall kiln during a normal carbonization cycle. In the kiln, the front moves from one end of the kiln to the other.

Fig. 7-5 (Top View)

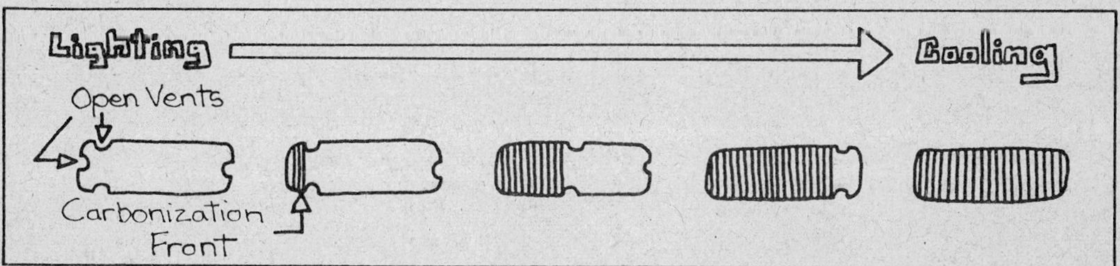

Observing the Progress of a Kiln

Now that the carbonization process is started it is important to be able to follow the progress so that controls may be operated at the proper time. In a few types of kilns the process will progress evenly with the entire charge going through each phase of carbonization at about the same time. However, in most kilns the wood will carbonize along a slow moving front.

Here are some signs which will indicate what is going on inside the kiln:

All Kilns

Ignition − − − − − − − − − − − − − − − − − *Carbonization complete* − − *cooling*

	Ignition		Carbonization complete		cooling
Type of smoke	Thick, dark	Thick, white, billowy as moisture is driven off	Thin, yellow as the wood starts to break down	Thicker, blue smoke as the charcoal starts to burn	None
Smell of smoke	Like burning wood		Alcohol and methane	Burning charcoal	None
Feel of smoke	Hot	Cool, moist	Hotter, oily	Hot	None

Temp. of the kiln surface	By touching the kiln and feeling where the heat is, you can find more where the front is within the kiln.
Sound of the kiln	As the wood carbonizes you will hear snaps and crackles in different parts of the kiln resulting from the shrinking of the charge as it becomes charcoal.
Time	After running a particular kiln a number of times you can make a reasonable guess how it is progressing just by knowing how long it's been since you lit it.

Earth Covered Kilns

Shape

As the charge carbonizes it will shrink causing the kiln to change its shape.

Color of the earth cover

The tars driven off the wood will darken the earth covering.

Inspection

If you are uncertain what a particular part of the kiln is up to, open up that section and find out.

Trouble Shooting

Very few runs are made without at least some difficulty. Here are some possible problems and solutions.

Collapse
(earth covered kilns)

This will occur less often if the kiln is tightly packed. Loose earth and vegetation should be stockpiled and tools should be at hand as the hole must be patched immediately or the charge will quickly start burning out of control (Fig. 7-6).

A hole can easily be spotted from the smoke that comes pouring out. If the hole is large it may be necessary to stuff it with additional wood to support the patching material. Cover the hole with vegetation and then earth.

If the kiln is large enough that it becomes necessary to climb on top to work, *step only on walkways of poles or boards* (Fig. 7-7). These distribute your weight over a greater area and lessen the chance of an accident.

Be very careful on top of a kiln!

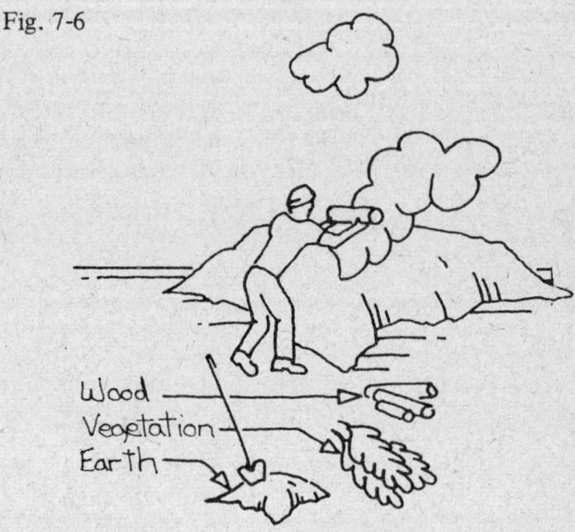

Fig. 7-6

Fig. 7-7

Cracking
(cement, adobe, masonry kilns)

Cracks are caused by differing expansion rates of the materials that make up the wall. These cracks will admit air and make the kiln difficult to control. Keep a lime plaster handy for patching and keep the cracks filled (Fig. 7-8).

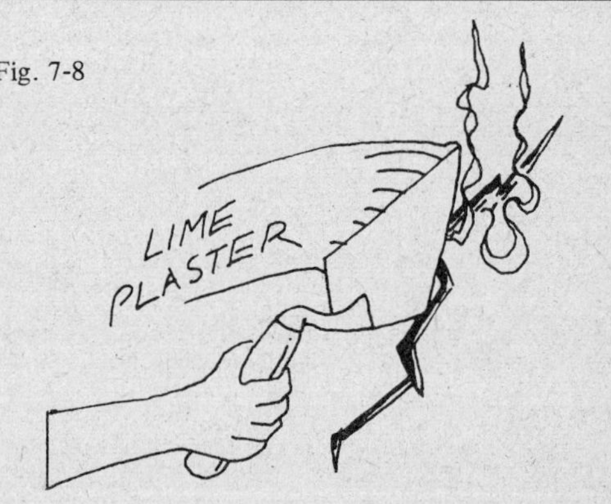

Fig. 7-8

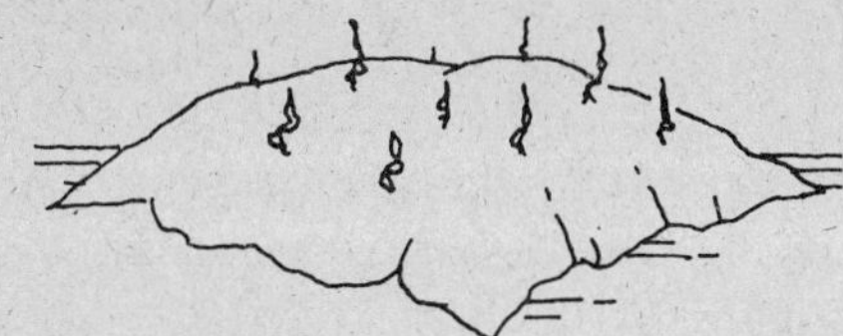

Small Smoke Leaks
(earth covered kilns)

As the run progresses there will be many small smoke leaks (Fig. 7-9). A shovel full of earth will stop the largest, and the rest are harmless.

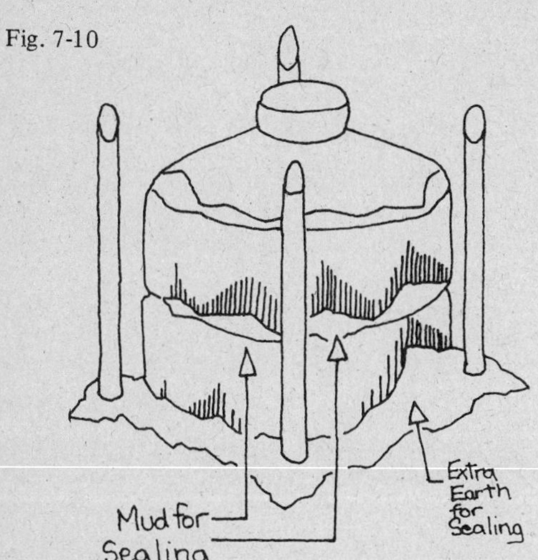

Warping
(metal kilns)

There may be difficulty in sealing the joint between sections of a metal kiln if warping has occurred. If the joint can't be sealed with sand, use mud and keep checking the mud for cracks. If the joint is not well sealed the air leak will cause a heat buildup and further warping.

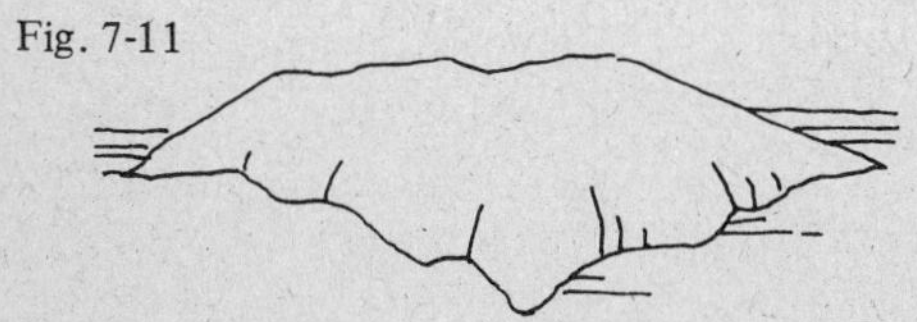

Slow Down

As already mentioned, slow down is expected after ignition, but if it happens later in the run open more vents. If the kiln still does not respond, open it and relight.

Carbonization Is Too Fast

Slow the process down by closing off vents. If there is no response, look for cracks.

Uneven Front Movement

Uneven movement of the carbonization front may result in portions of the charge being missed and not carbonized or incompletely carbonized. This may be caused by wind, improper vent operation, uneven packing, differing wood conditions. Opening vents in the problem area will help. If the problem is caused by wind, open lee vents (the vents sheltered from the wind) and close windward vents (the vents facing the wind).

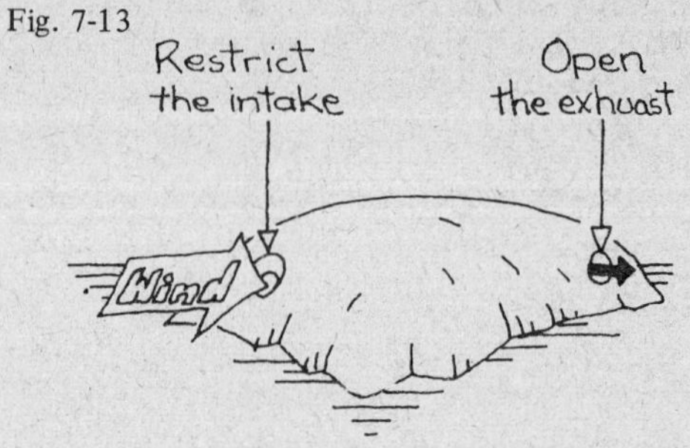
Fig. 7-13

Puffing

Wind will cause occasional puffs of smoke from one or more vents but if a vent starts blowing and sucking in a rhythmic fashion then a section of the charge is on fire. Quickly close all vents and check for cracks. After several minutes open one vent and check if the puffing has stopped. If so, gradually open other vents until operation is normal.

Fig. 7-14

Sealing the Kiln

When the charcoal is finished the smoke will turn a clear light blue and diminish in volume. Most of the vents are probably already closed. Shut any remaining open vents and seal any leaks. The kiln will start to cool down. The cooling period will vary depending on volume and weather conditions, but overnight is a minimum for all kilns.

Opening the Kiln

Before extracting the charcoal, clean the kiln site so there will be space to put the charcoal. Metal, pit, masonry, adobe, or any other kiln which will be bodily entered by workers *must* be opened and aired out before entering. Carbon monoxide and carbon dioxide will be present but oxygen will not.

Earth kilns should be opened one small section at a time. After pulling out charcoal

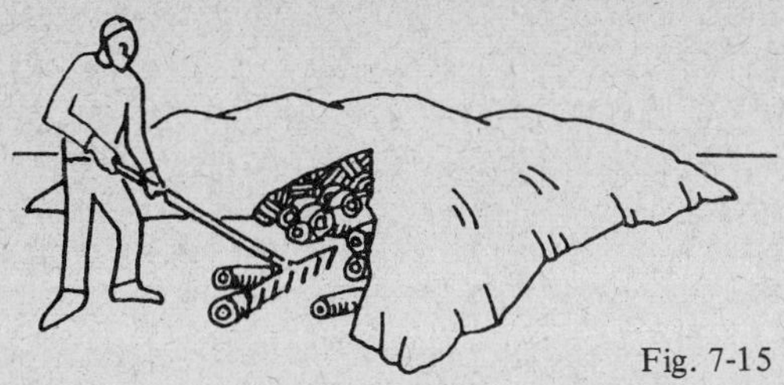

Fig. 7-15

since it is better to have wet charcoal than to have no charcoal.

After the charcoal has cooled it may be piled or loaded into jute bags (Fig. 7-17).

from one section recover the area with earth. If a flareup occurs cover the area immediately and let the kiln cool further (Fig. 7-15).

The charcoal should be spread out and observed for fire (Fig. 7-16). Fire should be extinguished with earth rather than water which diminishes the charcoal's quality. A serious fire can be extinguished with water

Fig. 7-16

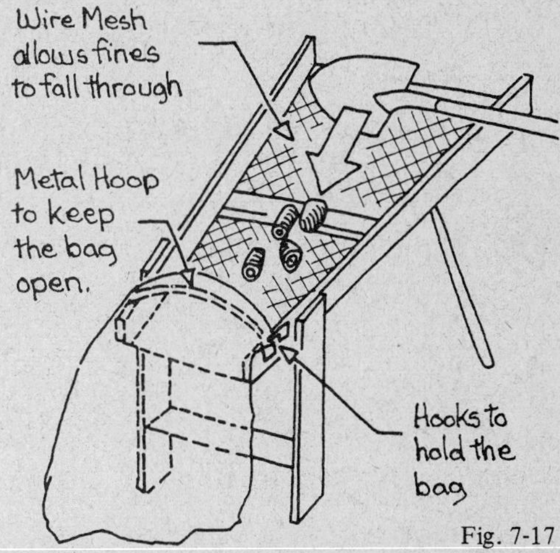

Fig. 7-17

8. Special Kiln Types

There are several types of kiln which are constructed or operated differently.

Continuous Kilns

In the simplest form a continuous kiln is a cylinder with a tight fitting lid, vents at the top and bottom, and a door at the lower end (Fig. 8-1).

To operate a continuous kiln a fire is started in the bottom, the kiln is filled with wood, and the top is put in place. The wood in the top of the kiln is heated by gases from the exothermic zone in the middle section. After carbonizing, it cools in the bottom of the kiln before removal from the bottom. As the charge shrinks and charcoal is removed, additional wood is added at the top. This procedure can be continued as long as material is available. Short, regular sized pieces of wood with the same moisture content enable even carbonization.

The type of kiln in Fig. 8-2 is constructed from metal. The exothermic zone may be lined with refractory material such as fire brick. There are usually various systems connected with this type of kiln for recovering heat and condensables. As a result of all of this the cost is high; however, simpler forms could be built locally.

Furnaces

Continuous kilns employing a mechanical means of moving the charge are called furnaces (Fig. 8-3). These kilns are useful

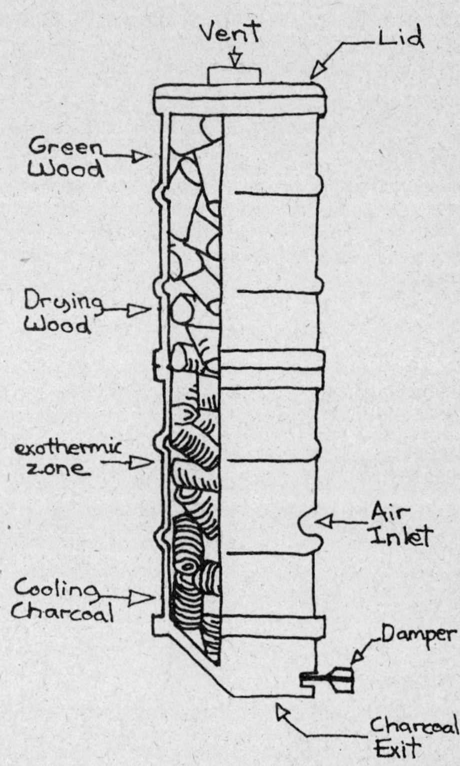

Fig. 8-1: Continuous Kiln

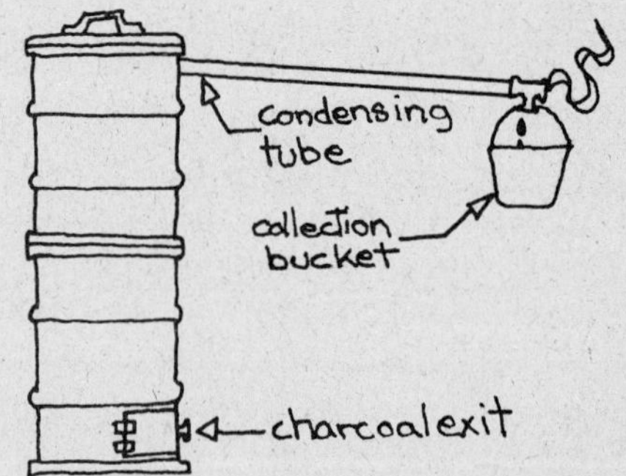

Fig. 8-2: Simple Condensable Collection

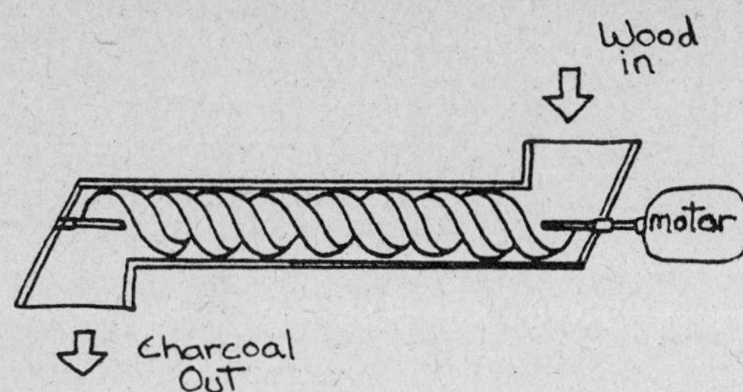

Fig. 8-3: Furnace

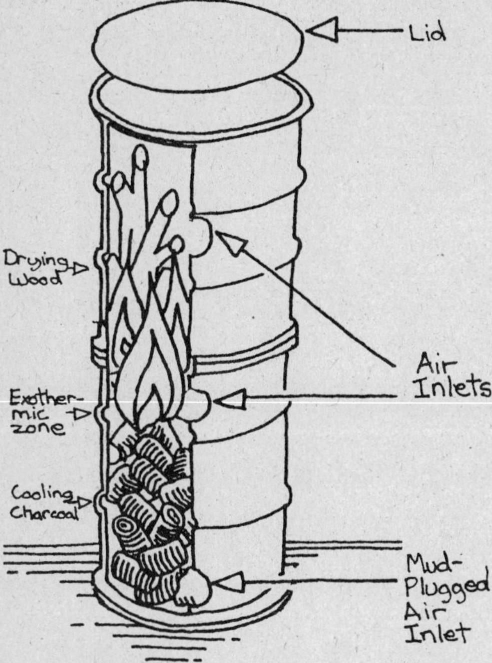

Fig. 8-4: Drum Kiln

Fig. 8-5: Retort

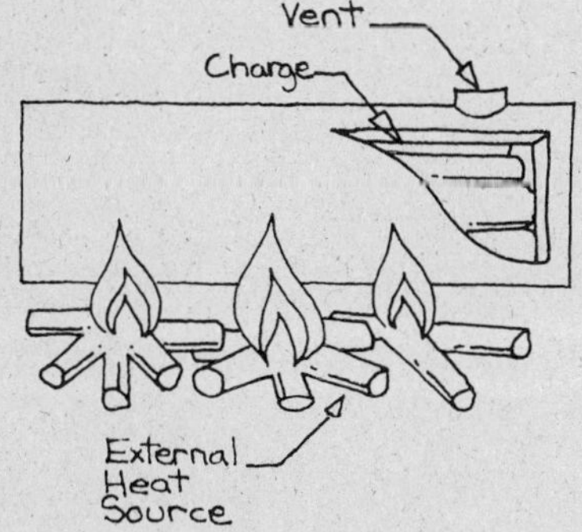

for carbonizing fine material such as sawdust which insulates itself and blocks air passage. The major disadvantages of furnaces for small scale production are cost and complexity.

CUSAB Type Kilns

A CUSAB kiln is a large, expensive, metal kiln; however, a simple type called the drum kiln has been developed (Fig. 8-4). It is constructed by welding two 220 liter oil drums end to end to form one long drum with a solid bottom. Three holes about 5 cm in diameter are spaced evenly in one side starting at the base. This kiln works very fast. It can be filled with charcoal in 2–3 hours.

The kiln operates similarly to the continuous kiln. A small fire is started in the bottom and wood is gradually added as it burns. There will be very little smoke unless wood is added too fast. The burning wood forms charcoal which falls to the bottom. When the bottom air hole is choked with coals it is sealed with mud. As there is no oxygen under the fire, the charcoal starts cooling. The process continues until all the vents are plugged and the kiln is full of charcoal. A lid is then placed on top and sealed with earth, and the kiln is cooled.

The drum kiln is very suitable to make coconut-shell-charcoal.

Retorts

A retort is a closed metal container to which the heat for carbonization is applied from outside.

Advantages
— Clean charcoal.
— Good heat control.
— No loss of charge to supply the heat.

Disadvantages
— Usually high cost.
— Most models are complex to build and
  operate.

Retorts in their simplest form (Fig. 8-5) are
very easy to operate. A metal box is loaded
with wood and sealed except for a small
exhaust vent. Heat is applied from outside
of the box until the charge is carbonized.
The charcoal is allowed to cool and then
removed.

To be economical the heat used to power a
retort should come from a surplus source
that is not normally used. Part of the heat
can be supplied by burning the gases produced in the retort.

9. Charcoal Types and Uses

There are numerous home, industrial and commercial uses for charcoal and each has its particular requirements. If you are hoping to produce charcoal for a specific use, be sure to thoroughly investigate the requirements first as this will directly influence raw material selection and conversion method and operation. A few examples are given below. Since there is a good market for activated charcoal we have also included a small amount of information on charcoal activation even though it is a high capital technology in its present form.

Home

Domestic cooking requires a charcoal which is not too hard to light but still puts out a good radiant smokeless heat and lasts a long time. These requirements can probably be met by a blend of charcoal from dense wood for long lasting heat and lighter wood for ease of ignition.

Small Foundries

Small scale foundries, such as those which cast aluminum pots from recycled metal, are good markets for fines which are not saleable for cooking fuel.

Steel Making

Different industrial uses require specific characteristics. For example, charcoal for steel making must have a carbon content of at least 80%, less than 1.5% ash, with low phosphorus content, low volatile content, no fines, and a high load bearing strength in order to stand up when mixed with ore and limestone in a steel furnace.

Artisans

General blacksmiths can also use fines, but jewelers usually need a more uniform and cleaner charcoal.

Cement Manufacture

Charcoal for use as a heat source in cement manufacture should have less than 4% moisture content and less than 4% ash, but a volatile content of 20% or more. This requirement, which ensures combustion as the powdered charcoal is blown into the furnace, means that the final carbonization temperature should be below 500 °C. Charcoal from earth covered kilns is usually not acceptable as it contains too much ash. Sawmill waste with its high percentage of bark is also not acceptable for the same reason.

Activated Charcoal

Activated charcoal is charcoal from which the hydrocarbon tars which adhere to the carbon have been removed. These tars have a high boiling point and are not removed in the normal carbonization process. Their removal increases the surface area of the charcoal about 2000 times, and creates a vast network of very small channels. This intricate structure greatly increases the filtration and adsorptive qualities. The process of activation adds greatly to the marketability of charcoal; however, it also increases the capital investment needed because of the more complex technology.

The most common method for activating charcoal is to heat it to $700° - 950$ °C in an atmosphere of superheated steam or carbon dioxide and hold it for a time ranging from 15 minutes to several hours. This process will reduce the carbon as well as the hydrocarbons, so eventually a point will be reached where further activation will result in too much loss of material.

Another method of activation is to mix the material to be carbonized with a chemical activator before carbonizing. Some possible activation chemicals are calcium chloride, magnesium chloride, zinc chloride, phosphoric acid, sulphuric acid, sodium phosphate, dolomite.

10. Kiln Design Evaluation

There are several ways to evaluate a kiln. Two will be discussed here.

Evaluation by User

One of the simplest evaluations is to observe if the kiln is accepted and used by the charcoal makers. If the kiln has been repeatedly used, then it may be assumed to be accepted. Once the charcoal maker is familiar with the kiln, it is a good time for an interview to find out what his likes and dislikes are. Any modifications he has made in the construction or operation of the kiln should be noted for evaluation.

Evaluation by Yield Calculation

Kiln yield figures are used to compare different kilns and evaluate changes in design or operation. When someone speaks of kilns yielding a certain percent this means the weight of charcoal produced compared to the weight (oven dry) of the wood used to produce it:

$$\frac{\text{Weight of Charcoal Produced}}{\text{Weight of Charge (Oven Dry)}} \times 100 = \text{Yield.}$$

It is immediately obvious that all of the wood in the kiln couldn't be oven dried before carbonizing, therefore this oven dry figure is a calculated figure based on air dry wood weight and wood moisture content.

To get exact yield measurements it is necessary to weigh all of the wood that goes into the kiln. Except for well funded research operations, this is not possible for large kilns. However, weight can be sampled and an average weight per stere figure can be arrived at for local use. Moisture content can be measured by weighing samples, then drying them to a stable weight and calculating the percent moisture in the wood using the following formula:

$$\frac{A - B}{B} \times 100 = \% \text{ Moisture}$$

where A is the wet weight, B is the oven dry weight.

From this and the total weight of the wood the oven dry weight of the charge may be calculated. For example, a 100 stere kiln has an average stere weight of 325 kg obtained by weighing 10 sample steres. 20 wood samples weigh 86 grams before drying and 63 grams after drying.

$$\frac{86 - 63}{63} \times 100 = 36.5\% \text{ Moisture}$$

Total wood weight is:

325 kg × 100 steres = 32500 kg.

Calculated oven dry weight is:

$$\frac{32500 \text{ kg}}{1.365} = 23809 \text{ kg}$$

(1 + 36.5% expressed as a decimal).

The total weight of charcoal produced was 6653 kg.

Kiln yield is:

$$\frac{6653 \text{ kg charcoal}}{23809 \text{ kg oven dry wood}} \times 100 = 27.9\% \text{ yield}$$

This is all well and good, but what do you do with the wood that didn't completely carbonize? These pieces are called brands. They can be used to evaluate the kiln's performance by noting how many you get and what part of the kiln they come from. As a general rule, 2% or less of the total kiln volume is an acceptable level of brands. This an arbitrary figure and may not apply to all kilns. Huge chunks of wood in the kiln may take several runs to totally carbonize. Brands can be considered as oven dry wood and subtracted from the total calculated oven dry wood weight. In the above example calculating out 2 steres of brands would give:

238 kg (average oven dry stere weight) $\times$ 2 = 476 kg

23809 kg − 476 kg = 23333 kg

$$\frac{6653 \text{ kg charcoal}}{23333 \text{ kg oven dry wood}} \times 100 = 28.5\% \text{ yield}$$

If it is a research kiln, weigh the brands for more accuracy.

This calculation bothers some people. It seems somehow like cheating by removing an obviously unsucessful part of the charge which then results in an increase in the yield. However, if you consider that the brands had no part in the final charcoal yield, i.e. they didn't end up as charcoal, then they don't belong in the calculation. This is why we establish a percentage cutoff point for brands. Above the cutoff there are too many brands indicating the kiln didn't work properly anyway so there is no need to bother with yield measurements.

It is more meaningful to state all of the variables and assumptions when quoting yield figures. For example, a typical statement might say: 100 stere kiln, 325 kg average stere, 36.5% moisture content oven dry basis, 476 kg brands, 28.5% yield brands excluded.

Kiln yield figures don't give much useful information to the forester who wants to know how much charcoal can be produced from land areas or solid wood volumes. A more useful figure might be kilograms of charcoal per stere of charge. Kilograms per stere is also more useful in monitoring a kiln once a design is in operation and you know what this figure should be for your conditions and the time of year. For example, if you are running a 100 stere Casamance Kiln in southern Senegal in the dry season, you know that this figure should be between 90 and 130 kg/stere. If it isn't then you go find out why. This same figure will not apply to any other kiln type, size, location, season etc. It is a local normal production figure. It doesn't take into account moisture content or wood density. One Casamance Kiln in Southern Senegal hit 200 kilograms per stere − but that was using dry, very dense wood in the dry season under optimum conditions with trainees (who may be much more exact since they don't want to make a mistake) under the instruction of an excellent teacher.

To determine the yield of a specific kiln design requires the evaluation of several runs. If there is a variation of more than a few percentage points in the results, more repetitions will be required.

11. Briquetting

Charcoal briquettes are made by compacting and binding small pieces of charcoal. These fines may be from carbonization of small material such as grass, sawdust, or peanut hulls, or they may be the residue of wood carbonizing. Although agricultural wastes can be carbonized they should be returned to the earth if possible.

Fig. 11-1

Advantages
— Utilizes fines.
— May utilize agricultural or industrial waste.
— Can make dense charcoal from light wood.
— Contains more energy per unit weight than regular charcoal.
— Slower release of heat.
— Standard size allows standard unit for marketing and use.
— More durable for transportation.

Disadvantages
— Needs promotion because is not a traditional fuel.
— More expensive due to:
1. special production equipment needed,
2. more production steps,
3. more middlemen,
4. longer production time,
5. more money in inventory.

Characteristics of a Good Briquette

A briquette needs to be strong enough not only to stay together, but be able to take rough handling in packaging and transport. The briquette should also retain its shape while burning.

Making Briquettes

To make briquettes you will need the following:
— Charcoal fines,
— Binder to stick the fines together,
— Carrier — a liquid to distribute the binder,
— Compaction device,

There are three methods of forming a briquette: by use of extreme pressure; some kind of binder; or a combination of the two.

To make briquettes by extreme pressure requires very expensive equipment while briquettes made with binders alone are weak and friable. In most cases a combination of the two is used. We will concentrate on the combination method in this section.

Binders

There are many substances that can be used as binders including starch, clay, animal manure, molasses, plant resins, and petroleum products.

In looking for a binder it is a good idea to ask people what they use to stick things together. Try a whole bunch of their ideas

to see which one works best. The qualities of a good binder are:
— low in cost,
— good binding ability,
— good smell while burning,
— readily available,
— soluble in a low cost carrier.
See Appendix D for starch extraction instruction.

Avoid binders that are used for food because you do not want to put additional pressure on the often limited food resources. Similarly, animal manures are best used as fertilizers. Clay lowers the heating value of the briquette, but it can be used in combination with another binder to extend the burning time of the briquette. When tar or molasses are used the briquettes must be heated to drive off the added volatiles. Petroleum products work best with petroleum carriers. Both can be expensive.

The briquetting process can be broken into four steps: grinding, mixing, compaction, and drying.

Grinding

The charcoal must be fine enough to form good briquettes. The size of particle will vary according to the type of charcoal and how the briquettes are to be made. Grinding is done by several methods. The simplest method is to use a large mortar and pestle. This is slow and requires a great amount of very hard work. Some grain grinders can be used if modified for more durability.

Mixing

There are many recipes for making briquettes. Here are a couple of formulas for guidance:
100 kg charcoal, 5—7 kg starch, 30—35 kg water;
100 kg charcoal, 15—30 kg tar, 1 kg fuel oil, 30 kg water.
The exact mixture used will depend on how fine the charcoal is ground, the type and amount of binder used, and how much compaction can be applied. Small batches should be made until the briquettes are satisfactory.
The most accessible place to mix is on hard packed earth, though this does cause a certain amount of contamination. Smaller batches can be done in drums, or you can use a concrete slab.

Compaction

Compaction determines the shape and density of the briquette. Size and shape will depend on the end use and the desire of the consumer. The simplest method is to use

Fig. 11-2

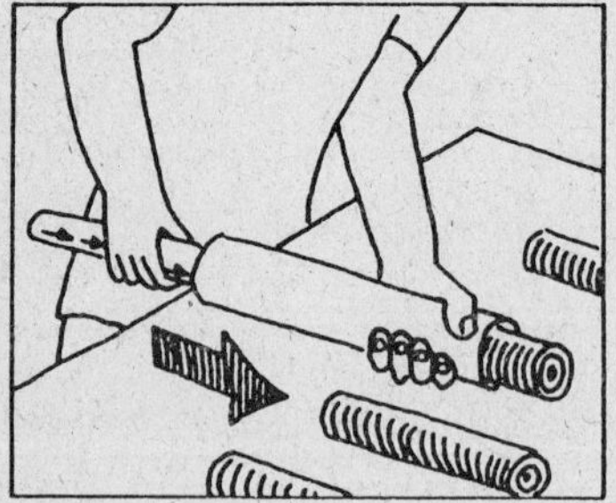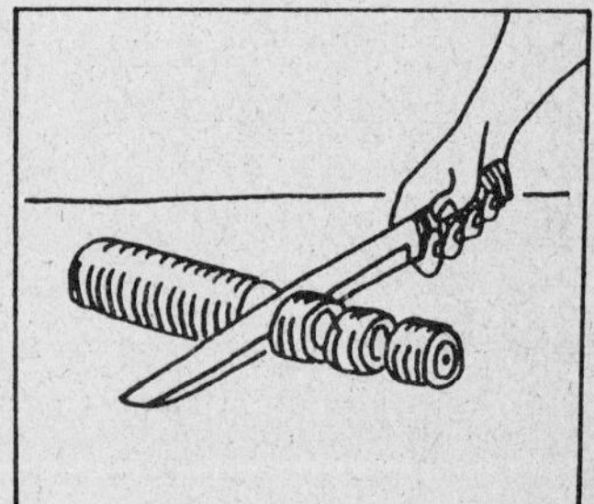

your hands to form and compress the briquette, but not much pressure can be applied with this method. The method shown in Fig. 11-2 works well if pipe or tubing is available.

An earth brick press (Fig. 11-3) will make a dense briquette when used with an insert to increase the compression ratio. Other means of getting increased pressure include screws, weights, and hydraulics. There are also several commercial high pressure briquette presses available, but they are too costly and complicated for small scale production.

Drying

The simplest, cheapest method for drying is to just lay them out in the sun. A solar dryer can speed the process considerably. Waste heat from a retort can be harnessed for drying briquettes but internal dryer temperature must remain below briquette ignition temperature. Drying time will depend on the amount of moisture in the mix and the type of dryer used.

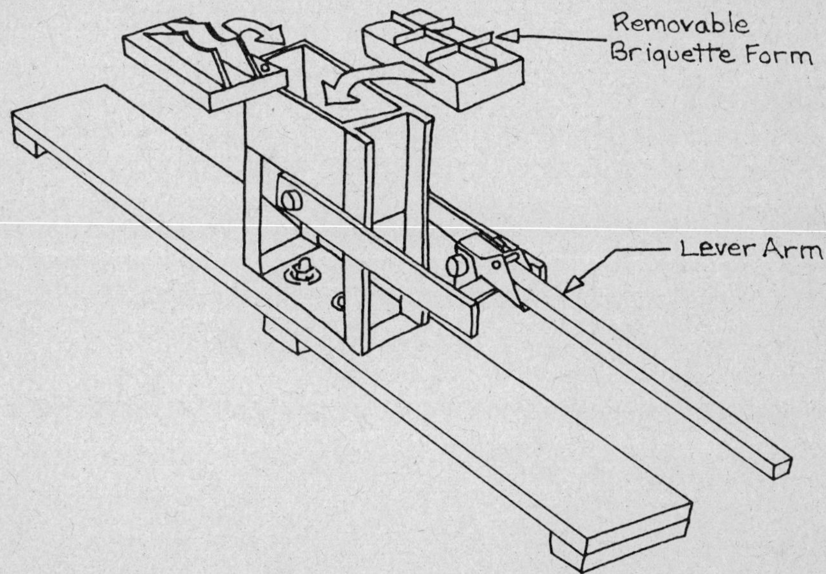

Fig. 11-3: Earth Brick Press for Briquetting (for more information: CICON, Centro de Investigaciones Ingenieria, Ciudad Universitaria, Zona 12, Guatemala, Central America)

12. Charcoal Stoves

Fig. 12-1

The use of improved charcoal kilns is very important, but of equal importance is using efficient charcoal stoves. As we have seen, the combination of efficient kilns and improved charcoal stoves can be quite good as far as overall energy efficiency is concerned.

It is essential that any program of charcoal stove extension work should include the following elements:

1. an assessment of needs, resources, and conditions in the area;

2. a design process that includes the builders and eventual users of the stove;

3. an evaluation of fuel savings, user acceptability, innovations, and problems in the stove and in the stove program.

Stove Design Considerations

An efficient charcoal stove is one which transfers as much of the heat as possible from the fuel to the cooking pot. In order to better understand this process let us define the three ways heat is transferred (Fig. 12-2).

Radiation is the transfer of energy by radiant waves, much like light waves which move in straight lines, in all directions equally. A practical example is heat from the sun.

Conduction is the transfer of heat through solid objects. Touching a hot stove with your hand is a good example.

Convection is the transfer of heat through the movement of a gas or liquid. Hot gas rising up a chimney is an example of this.

Fig. 12-2

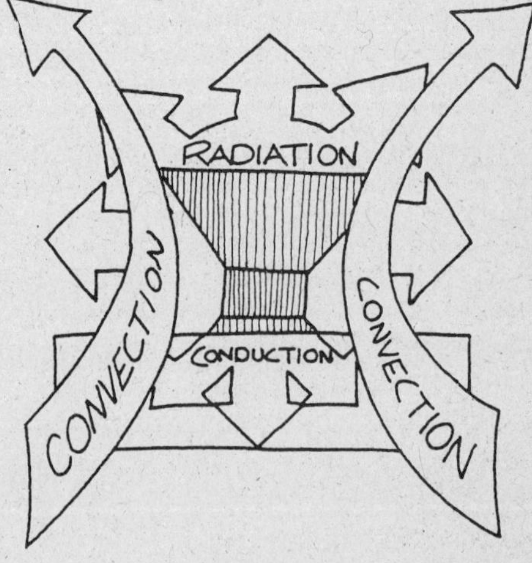

Listed below is a series of guidelines to keep in mind when designing a charcoal stove.

1. *Charcoal transfers heat mainly through radiation.* Due to the lack of volatiles in charcoal, very little convective heat is produced. This leads to two design guidelines:

a. The pot should be placed directly on or very close to the charcoal. By doing this you will maximize radiant heat transfer (Fig. 12-3).

b. Only a pot directly over the charcoal receives a significant amount of heat. There is only a very small amount of hot gases to heat a second pot.

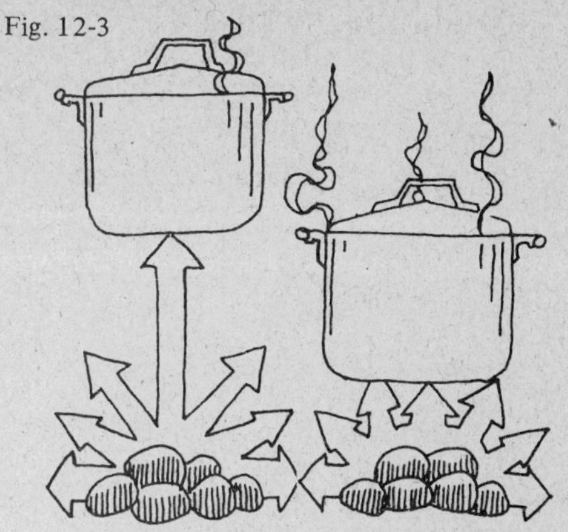

Fig. 12-3

2. *Some means of igniting the charcoal should be provided.* Although some people start charcoal with kerosene, it is good to have a small space under the charcoal bed where a fire can be lit (Fig. 12-4).

Fig. 12-4

3. *A grate of 25–35% free area is recommended* (Fig. 12-5). Though a few charcoal stoves don't use them, grates are recommended to promote even burning and to keep the charcoal from becoming covered with insulating ash. More free area may be needed at high altitudes.

Fig. 12-5: 25% free vent area means 1/4 of the grate is holes

4. *Air should enter through the bottom of the grate* (Fig. 12-6). It is also a good idea to have some method of shaking the grate so that the ash falls through. The size of the holes in the grate will depend on the size of charcoal being used.

Fig. 12-6

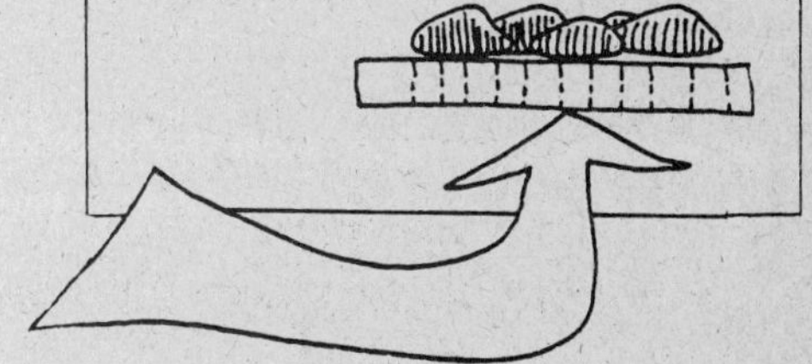

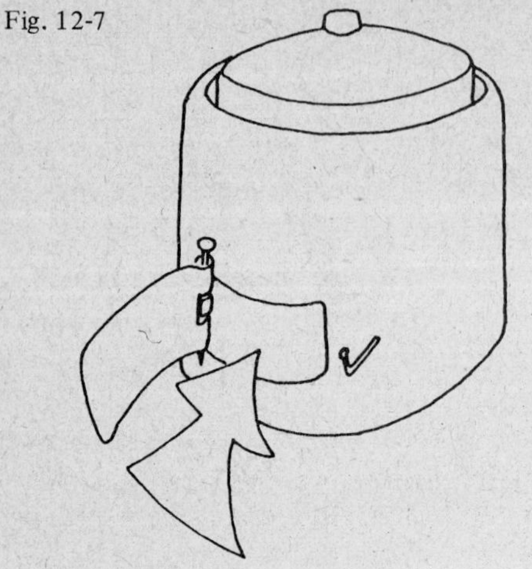

5. *Some kind of air inlet control is helpful, especially for longer cooking.* The damper in Fig. 12-7 restricts the amount of air reaching the charcoal, which controls the rate at which the charcoal burns.

6. *The pot and charcoal should be protected from wind and convective air currents* (Fig. 12-8). These air currents can take away a great amount of heat. An added advantage of wind shields is that any heat not absorbed by the pot bottom can be directed up along the pot sides.

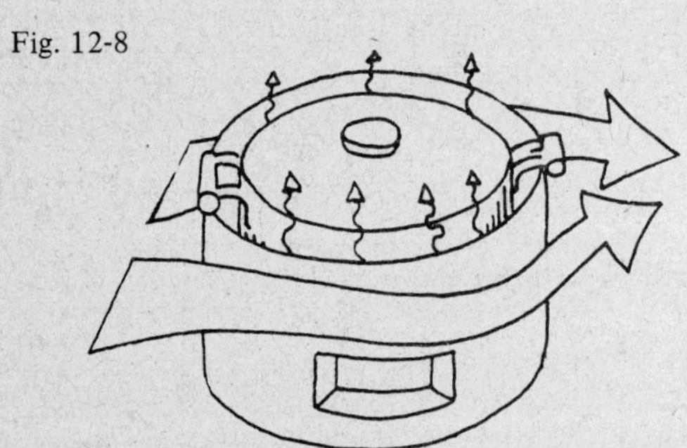

7. *Charcoal stoves should be insulated to lessen conductive heat losses* (Fig. 12-9). Low cost materials such as ash and sand work well for this purpose. The wind shields mentioned above can also be insulated to further protect the pot. In the simplest of stoves the unburned charcoal insulates the burning charcoal.

8. *The stove should be stable with the pot in place.* Many charcoal stoves have a small base and when used with a big heavy pot can be very dangerous (Fig. 12-10).

A note of caution:
Remember that charcoal should be used only in well ventilated areas (Fig. 12-11). Charcoal is smokeless, but is far from gas-less. The fumes can be deadly in enclosed areas.

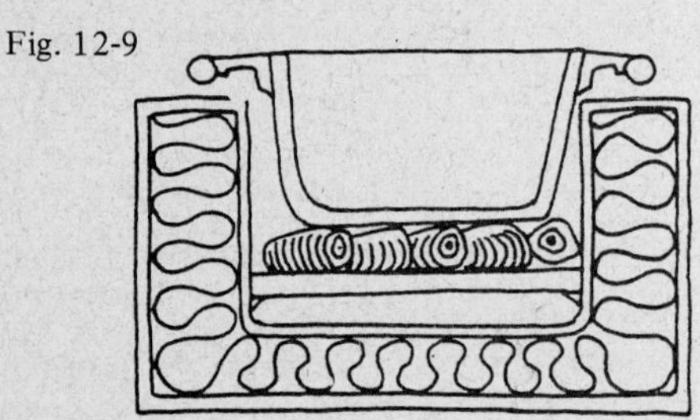

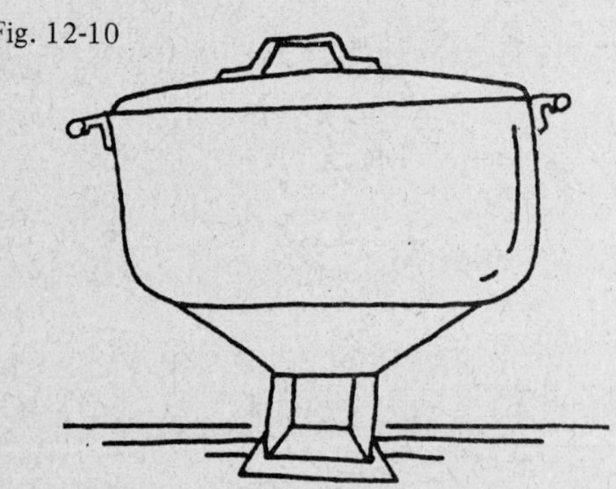

Fig. 12-11

A Review of Charcoal Stoves

In this section we will look at several different stoves in use around the world. It is important to keep in mind that these are adapted to a particular set of conditions and should not be looked on as the perfect solution for your situation. More than likely a totally new stove or a modification of one of these designs will be needed to fit the conditions in your area.

Fourneau Malgache

The Fourneau Malgache (Fig. 12-12) is a simple stove made from sheet metal. This type of stove is used in Central America and West Africa.

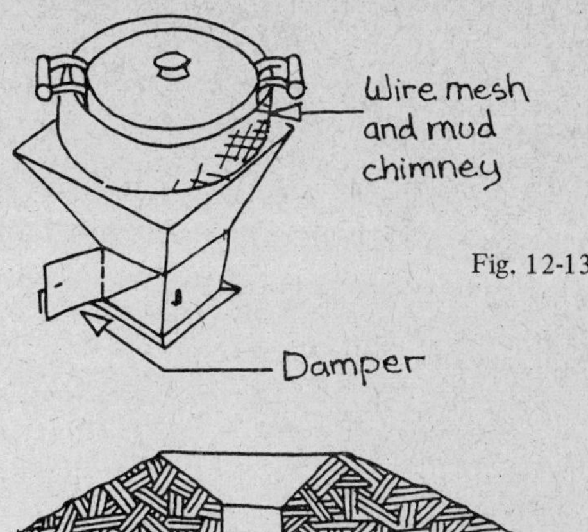

Fig. 12-13

Fig. 12-14: Sand Insulation

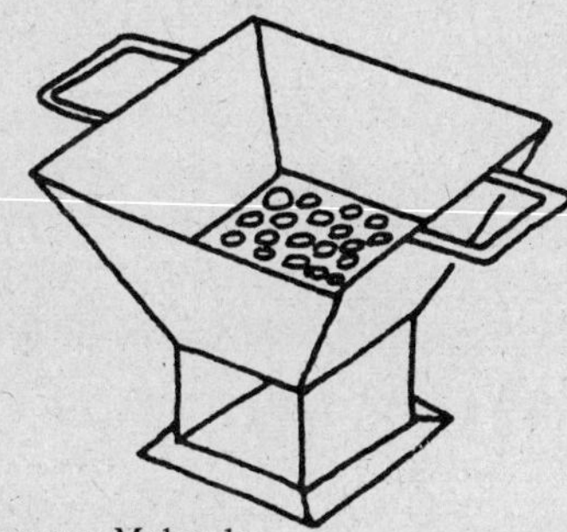

Fig. 12-12: Fourneau Malgache

Advantages
— Low cost.
— Simple.
— Local construction.
— Portable.

Disadvantages
— Unstable.
— High heat loss.

Improvements
— Damper to control draft (Fig. 12-13).
— Wind shield.
— Insulate the sides.
— Heap sand around the stove for insulation and stability (Fig. 12-14).

Bowl Stove

The bowl stove (Fig. 12-15) is a traditional ceramic stove used in West Africa. To stabilize it a shallow hole is dug and the stove placed inside.

Fig. 12-15: Bowl Stove

Advantages
— Portable.
— Stable.
— Local construction.
— Low cost.

Disadvantages
— May not be very efficient.
— Poor air circulation.

Improvements
— Use a grate.
— Collar around the pot for wind protection.

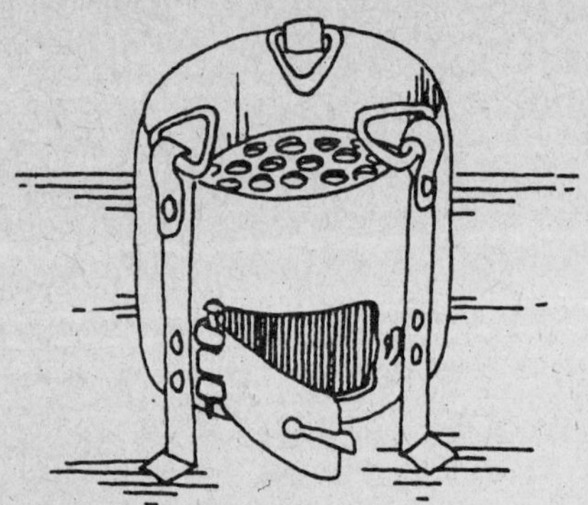
Fig. 12-16: Jiko Stove

Jiko

The Jiko (Fig. 12-16) is a metal stove used in East Africa.

Advantages
— Effective damper (when used).
— Local construction.

Disadvantage
— Radiates heat.

Improvements
— Add a clay/sand liner to the top section.
— Some work has been done on commercial ceramic liners for the Jiko.

Terra Cotta Stove

A ceramic stove found mostly in Asia. Terra cotta as a material is more durable than unfired clay, yet light in weight (Fig. 12-17).

Advantages
— Cheaper than metal stoves.
— Less heat loss than metal stoves.
— Traditional skills.
— Portable.
— Local construction.

Disadvantage
— Fragile.

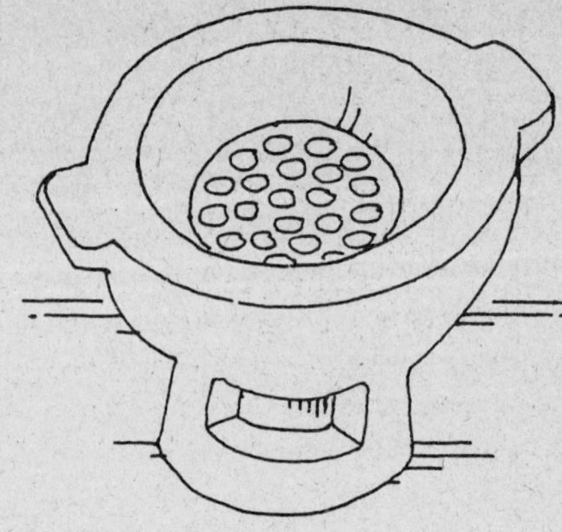
Fig. 12-17: Ceramic Stove

Improvement
— Add a damper to control draft.

Thai Bucket

The Thai Bucket (Fig. 12-18) is a very efficient stove made of metal and ceramic used in South East Asia.

Advantages
— Well insulated.
— Good damper system.

Disadvantages
— Complicated to build.

Improvement
— Widen throat to allow pot to fit down inside stove.

Fig. 12-18: Thai Bucket

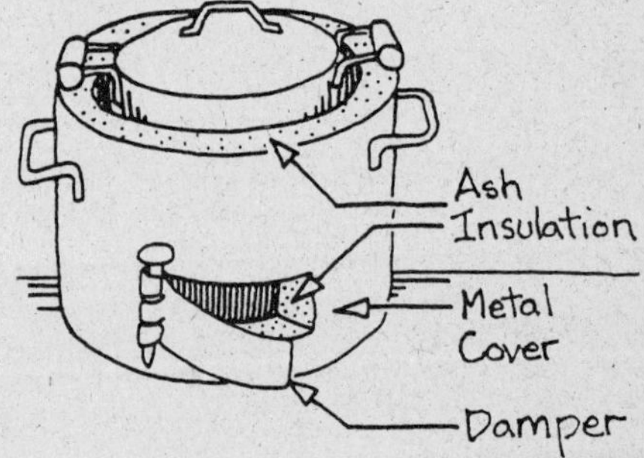

Appendix

A: Tables

Table A

Charcoal Yield Efficiencies
1 kilo wood 20% moisture

Charcoal Use Efficiencies
energy used for various charcoal stove efficiencies

A	B	C	D	E	F	G						
oven dry weight (kg)	yield %	yield kg charcoal	kcal/kg charcoal	kcal/kg yield	avail. energy %	loss %	energy available from column E					
							20%	25%	30%	35%	40%	45%
.83	10	.08	7100	592	16	84	118	148	178	207	237	266
.83	11	.09	7100	651	17	83	130	163	195	228	260	293
.83	12	.10	7100	710	19	81	142	178	213	249	284	320
.83	13	.11	7100	769	20	80	154	192	231	269	308	346
.83	14	.12	7100	828	22	78	166	207	249	290	331	373
.83	15	.13	7100	888	23	77	178	222	266	311	355	399
.83	16	.13	7100	947	25	75	189	237	284	331	379	426
.83	17	.14	7100	1006	26	74	201	251	302	352	402	453
.83	18	.15	7100	1065	28	72	213	266	320	373	426	479
.83	19	.16	7100	1124	30	70	225	281	337	393	450	506
.83	20	.17	7100	1183	31	69	237	296	355	414	473	533
.83	21	..18	7100	1243	33	67	249	311	373	435	497	559
.83	22	.18	7100	1302	34	66	260	325	390	456	521	586
.83	23	.19	7100	1361	36	64	272	340	408	476	544	612
.83	24	.20	7100	1420	37	63	284	355	426	497	568	639
.83	25	.21	7100	1479	39	61	296	370	444	518	592	666
.83	26	.22	7100	1538	40	60	308	385	462	538	615	692
.83	27	.23	7100	1598	42	58	320	399	479	559	639	719
.83	28	.23	7100	1657	44	56	331	414	497	580	663	745
.83	29	.24	7100	1716	45	55	343	429	515	601	686	772
.83	30	.25	7100	1775	47	53	355	444	533	621	710	799
.83	31	.26	7100	1834	48	52	367	459	550	642	734	825
.83	32	.27	7100	1893	50	50	379	473	568	663	757	852
.83	33	.28	7100	1953	51	49	391	488	586	683	781	879
.83	34	.28	7100	2012	53	47	402	503	603	704	805	905
.83	35	.29	7100	2071	54	46	414	518	621	725	828	932

Stove Efficiencies

1 kg air dry wood 20% moisture content 3441 kcal

effic. %	kcal/kg used
2	69
4	138
6	206
8	275
10	344
12	413
14	482
16	551
18	619
20	688
22	757
24	826
26	895
28	963
30	1032
32	1101
34	1170
36	1239
38	1308
40	1376
42	1445
44	1514
46	1583
48	1652
50	1721
52	1789
54	1858
56	1927
58	1996
60	2065

Table B

Economics of Transport Wood vs. Charcoal
Wood cost: $ 4.00/stere, Charcoal production $ 150/Tonne,
3 stere = 1 T, 1 Tonne wood at 25% mc = $3.268 \cdot 10^6$ kcal,
1 Tonne charcoal = $7.100 \cdot 10^6$ kcal.

km	cost of hauling 10 Tonnen	cost of hauling 1 0000000 kcal		haul + production cost	
		wood	charcoal	wood	charcoal
5	6	2	1	5	22
10	11	3	2	7	23
15	17	5	2	9	23
20	22	7	3	10	24
25	28	8	4	12	25
30	33	10	5	14	26
35	39	12	5	16	27
40	44	14	6	17	27
45	50	15	7	19	28
50	56	17	8	21	29
55	61	19	9	22	30
60	67	20	9	24	31
65	72	22	10	26	31
70	78	24	11	27	32
75	83	25	12	29	33
80	89	27	13	31	34
85	94	29	13	33	34
90	100	31	14	34	35
95	106	32	15	36	36
100	111	34	16	38	37
105	117	36	16	39	38
110	122	37	17	41	38
115	128	39	18	43	39
120	133	41	19	44	40
125	139	42	20	46	41
130	144	44	20	48	41
135	150	46	21	50	42
140	156	48	22	51	43
145	161	49	23	53	44
150	167	51	23	55	45
155	172	53	24	56	45
160	178	54	25	58	46
165	183	56	26	60	47
170	189	58	27	61	48
175	194	59	27	63	49
180	200	61	28	65	49
185	206	63	29	67	50
190	211	65	30	68	51
195	217	66	31	70	52
200	222	68	31	72	52
205	228	70	32	73	53
210	233	71	33	75	54
215	239	73	34	77	55
220	244	75	34	78	56
225	250	76	35	80	56
230	256	78	36	82	57
235	261	80	37	84	58

56

1. adobe — mud construction, usually in the form of blocks. Adobe is made of a mixture of clay soil, water, and sometimes straw.
2. brands — partially carbonized wood.
3. carbonization — the process by which organic matter becomes charcoal.
4. carbonization front — that area of the kiln in front of which is wood or brands, in back of which is charcoal.
5. charcoal yield efficiency — a figure that compares the weight of the wood loaded into a kiln to the weight of the charcoal which is withdrawn (see chapter 10).
6. charge — the organic matter to be carbonized.
7. fines — small particles of charcoal.
8. flareup — charcoal igniting if it has been removed from the kiln too soon.
9. logging slash — that wood that remains after an area has been logged.
10. overall efficiency — the percentage of energy available in the fuel that reaches the food in the cookpot. taking into account loses not only in use, but also in production. This figure is based on the best current test data available and may not be an accurate representation for what actually occurs. For more accurate information comprehensive, long-term surveys need to be done.
11. stere — one cubic meter of stacked wood.
12. stove efficiency — the percentage of the total energy available in a fuel that actually reaches the food in a cookpot. This figure may not be an accurate representation of what actually occurs in the kitchen. For more accurate information comprehensive, long-term surveys need to be done (see bibliography, *Stove Testing Procedures*).

Conversion Table

1 Meter		3.28 feet
1 Liter		0.2642 gallons
1 Kilocalorie	1000 calories	4185 Joules
1 Kilogram	1000 grams	2.2046 pounds
1 Kilometer		0.621 miles
1 Tonne	1000 Kilograms	1.1023 Tons (2204 pounds)
1 Stere	1 stacked meter cube	0.273 cords

C: Needs for Research

1. Large land clearance projects are often encountered in which natural forest is being cleared either for conversion to agricultural use or to forest plantations. The resulting woody material is usually windrowed with a crawler tractor and burned after drying. The planting schedule requires that it be cleared by planting season and the necessary time for cutting and piling for carbonization doesn't exist. The scale is such that enough metal kilns to handle the volume represents an impossible capital investment. The situation calls for use of the existing equipment to pile and cover the slash in some manner so it can be carbonized, thus preventing the waste of this resource. Yields would probably be fairly low but any return would be a gain over present methods.

2. Is the miniature kiln representative of full scale kilns? It appears to work like the full sized kiln. Carbonization times are not in scale and may be close to actual times. Gas pressures are a constant and will not change by reduction of size. Combustion temperatures are another constant. The miniatures appear to be more difficult to run in some designs, probably from reduction of heat mass. The big question is: are the yields comparable to full scale kilns? If so, miniature kilns will be an invaluable research tool. If a scale of comparison could be worked out, both teaching and research time could be dramatically reduced because miniature kilns allow the production output of approximately one kiln a day compared to one week or one month for full scale kilns. An added plus would be a reduction of equipment, labor, wood, and space requirements for research.

3. Further work is needed on briquetting machines, charcoal grinders, and mixers.

4. Accurate and complete long term energy surveys are needed to assess the impact of changes in fuel use.

5. What are the effects of insulation on a kiln? Could the heat be retained until the exothermic stage and then released? How would this affect yields?

6. Is there a simple method for refining of condensed volatiles?

7. Is there an inexpensive method for activating charcoal?

D: Starch Preparation

Most starches can either be purchased or are available free, but in most cases they must be extracted from the source. Starch is available from a wide variety of plants.

To extract starch, first grind or crush the plant material. The finer it is ground, the easier it will be to extract the starch. This process can be done with a hand grinder or with mortar and pestle. When greater quantities are needed, a pedal grinder may be the answer.

In the next step the starch is washed out of the grindings by soaking or rinsing. Don't dump the water — this is what you save. In order for the starch and water combination to work in the briquetting process it must be heated to 65 $^\circ$C to make it sticky. This can be done either before or after the briquette is formed. There must be the proper ratio of starch and water which is 5 parts water to 1 part starch by weight. Too much starch and the binder will be too stiff to work with, too little and it won't be strong enough. Here are two methods for achieving the right ratio.

1. The first method is to just try it and see if it works. Mix a small batch, compress it and see what kind of briquette that particular starch and water combination will make. Although the briquette will gain strength as it dries, it will give a good indication of strength while still wet. If the briquette seems too weak, let the starch suspension sit for three hours. The starch will settle to the bottom and the clear water on top can be decanted. This will concentrate the starch.

If the starch turns into a solid mass of gelatin when heated, add some water to the next batch. This method may seem slow, but as you become familiar with starch suspensions it will go more quickly. A good estimate based on experience can save a lot of time.

2. A more accurate method of doing this task is to stir the suspension, then take a sample and weigh it. As starch is 50% heavier than water, the sample will be heavier than a like volume of water. To determine how much starch is present use the following formula:

$(W - V) \times 2 =$ grams of starch per volume
W is the sample weight in grams
V is the sample volume in milliliters

Example:
100 ml sample weighs 110 grams
$(110 - 100) \times 2 = 20$
There are 20 gr of starch per 100 ml of suspension, so there are 90 gr of water (110 gr minus 20 gr) per 100 ml.

To establish the proper ratio of 5 water to 1 starch you would add 10 ml of water making the mixture 100 gr of water to 20 gr of starch. Correct the large batch by using a like proportion. If there is too much water, let the starch settle and decant.

Bibliography

Aprovecho Institute, 1984. *Fuel-Saving Cookstoves.* Deutsches Zentrum für Entwicklungstechnologien — GATE, Eschborn, Federal Republic of Germany.

Aprovecho Institute, 1980. *Guidelines for Evaluating the Fuel Consumption of Improved Cookstoves.* Africa Bureau, USAID, Washington D.C., U.S.A.

Arnold, John H. Jr., 1980. *State of the Art Review; Designing Rural Cook-Stoves.* Bioresources for Energy Project, U.S. Dept. of Agriculture and Forestry, USAID Washington, D.C., U.S.A.

Amalfitano, Gina and Cappaert, David, 1980. *Notes for Feu Malgache Tests.* Aprovecho Institute, Eugene, OR, U.S.A.

CILSS, 1978. *Energy in the Development Strategy of the Sahel.* Club du Sahel, Paris, France.

Conners, Stephen, 1982. *Wood Consumption in "Improved" Cookstoves.* Peace Corps, Benin.

Downey, John M., 1983. *Stove Testing Procedures.* Volunteers in Technical Assistance, Arlington, VA, U.S.A.

Earl, D.E., 1974. *Charcoal: an Andre Mayer Fellowship Report.* FAO, Rome, Italy.

Earl, D.E., 1975. *Forest Energy and Economic Development.* Clarendon Press, Oxford, England.

FAO, 1956. *Charcoal from Portable Kilns and Fixed Installations.* FAO, Rome, Italy.

Geller, Howard, 1981. *Cooking in the Ungra Area: Fuel Efficiencies, Energy Losses, and Opportunities for Reduced Firewood Consumption.* Centre for the Application of Science and Technology to Rural Areas. Indian Institute of Science, Bangalore, India.

Humphreys, F.R. and G.E. Ironside, 1977. *Charcoal from New South Wales Species of Timber.* Technical Paper No. 23, Forestry Commission of New South Wales, Sydney, Australia.

International Labor Office, 1975. *Charcoal Making for Small Scale Enterprises.* Geneva, Switzerland.

Karch, G.E., 1981. *Charcoal Stove Design and Test Results.* Food and Agriculture Organization of the United Nations (F.A.O.), Rome, Italy.

Karch, G.E., 1980. *Final Report of Forest Energy Specialist.* Project SEN/78/002, Ziguinchor, Senegal.

Karch, G.E., 1981. *Forest Energy in Cameroun.* FAO, Rome, Italy.

Karch, G.E., M. Wilburn, M. Wilburn, M. Boutette, 1982. *Small Scale Charcoal Making; A Manual for Trainers.* The Peace Corps Energy Project, Washington, D.C., U.S.A.

N'Diaye, I., 1980. *La Meule Casamancaise.* Project SEN/78/002, Ziguinchor, Senegal.

Paddon, A.R., A.P. Harker, 1979. *The Production of Charcol in a Portable Metal Kiln.* Tropical Products Institute, Ministry of Overseas Development, London, England.

Roos, Werner and Ursula, 1979. Revised edition 1984. *Construction of simple kiln systems.* Deutsches Zentrum für Entwicklungstechnologien-GATE, Eschborn, Federal Republic of Germany.

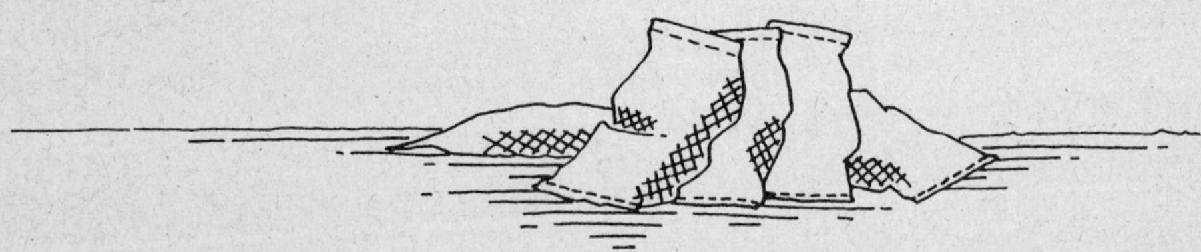